Non-Fossil Energy Development in China

Goals and Challenges

Non-Fossil Energy Development in China
Goals and Challenges

Edited by

Yunzhou ZHANG

Fuqiang Zhang, Shengyu Wu,

Bo Yuan, Junshu Feng,

Guanjun Fu, Nan Li, Libin Chen

ACADEMIC PRESS
An imprint of Elsevier

中国电力出版社
CHINA ELECTRIC POWER PRESS

Academic Press is an imprint of Elsevier
125 London Wall, London EC2Y 5AS, United Kingdom
525 B Street, Suite 1650, San Diego, CA 92101, United States
50 Hampshire Street, 5th Floor, Cambridge, MA 02139, United States
The Boulevard, Langford Lane, Kidlington, Oxford OX5 1GB, United Kingdom

Library of Congress Cataloging-in-Publication Data
A catalog record for this book is available from the Library of Congress

British Library Cataloguing-in-Publication Data
A catalogue record for this book is available from the British Library
ISBN: 978-0-12-813106-0

For information on all Academic Press publications
visit our website at https://www.elsevier.com/books-and-journals

Working together
to grow libraries in
developing countries

www.elsevier.com • www.bookaid.org

Publisher: Glyn Jones
Acquisition Editor: Glyn Jones
Editorial Project Manager: Naomi Robertson
Production Project Manager: Sruthi Satheesh
Cover Designer: Christian Bilbow

Typeset by SPi Global, India

Contents

Foreword ix
Preface xi
Introduction..xv

CHAPTER 1   Review and outlook of world energy development......1
    1.1 History and Status Quo of Development 2
        1.1.1 Population, Economy, Environment, and Energy
               Consumption 2
        1.1.2 Structure of World Primary and End-User Energy
               Consumption Structure (EUECS) 5
        1.1.3 World Energy Consumption and Production by
               Regions 6
    1.2 Energy Development Strategy and Enlightenment From
        Principal Developed Countries...10
        1.2.1 The US Energy Independence Strategy 10
        1.2.2 EU Energy Strategy 2020, 2030, and 2050 14
        1.2.3 Adjustment of Japan's Energy Strategy............................ 16
    1.3 Development Experience and Trend of World Clean Energy.....18
        1.3.1 Development Experience.. 19
        1.3.2 Development Trend 23
    1.4 Future Development Features of World Energy and Power
        System 31
        1.4.1 Features of Energy Development....................................... 31
        1.4.2 Development Trend of Electric Power 32

**CHAPTER 2   China's current situation of energy development
             and thinking on future development** 37
    2.1 Current Energy Structure and Non-fossil Energy Utilization
        in China..37
        2.1.1 Current Primary Energy Structure and Non-fossil
               Energy Utilization... 37
        2.1.2 Structure of Power-Generating Energy and Utilization
               of Non-fossil Energy 40
        2.1.3 End-User Energy Consumption Structure and Non-fossil
               Energy Utilization... 42
    2.2 Main Problems in China's Energy Development 44
        2.2.1 Rapidly Growing but Inefficient Energy Consumption..... 44

2.2.2 Energy Supply Security Is Insufficient 46

2.2.3 Irrational Energy Configuration 48

2.2.4 The Situation of Energy and Environment Is Grim 50

2.3 Ideas About Future Energy Development in China.....................54

2.3.1 Control of Total Energy Consumption (CTEC) 54

2.3.2 Energy Structure Adjustment and Layout Optimization ... 55

2.3.3 Construction of a Comprehensive Energy Transportation
System... 59

**CHAPTER 3   Methods for and models of research on non-fossil
energy development** **63**

3.1 Overall Analysis Thought and Method 63

3.2 Methods for Overall Optimization and Planning
of the Non-fossil Energy and Power System 66

3.2.1 Overall Framework of the Research Method.................... 67

3.2.2 Improving the Solution-Seeking Process 69

3.3 Model for Evaluation of Energy and Power Development 73

3.3.1 CGE Model.. 73

3.3.2 The IO Model 77

3.3.3 Energy Development Quality Evaluation
Indicator System 79

**CHAPTER 4   Conditions and development potential of energy
resources** **81**

4.1 Non-fossil Energy 81

4.1.1 Hydroenergy 81

4.1.2 Nuclear Energy 90

4.1.3 Wind Energy.. 97

4.1.4 Solar Energy 105

4.1.5 Biomass Energy 109

4.2 Fossil Energy 114

4.2.1 Coal.. 114

4.2.2 Oil 118

4.2.3 Natural Gas 122

**CHAPTER 5   Construction and comprehensive evaluation
of non-fossil energy development scenarios**...........**127**

5.1 The Main Principles of Scenario Analysis...............................127

5.2 Main Boundary Conditions...128

5.2.1 Predictions on Energy Demands 128

5.2.2 Power Demand.. 131
5.2.3 Non-fossil Energy Development Scale 135
5.2.4 Total Supply of Non-fossil Energy 138
5.3 Development Scenario Design 138
5.3.1 General Idea.. 138
5.3.2 Four Types of Scenarios... 141
5.4 Analysis on Scenario Optimization 148
5.4.1 Scenario 1: Expected Development Scenario.................. 148
5.4.2 Scenario 2: Hydropower Slowdown Scenario 165
5.4.3 Scenario 3: Nuclear Power Slowdown Scenario 173
5.4.4 Scenario 4: Demand Slowdown Scenario........................ 178
5.5 Comprehensive Evaluations of Scenarios 185
5.5.1 Proportion and Composition of Non-fossil Energy
Under Different Scenarios... 185
5.5.2 Benefits and Costs of Non-fossil Energy Development
Under Different Scenarios... 188
5.5.3 Basic Conclusions of Scenario Analysis......................... 190

CHAPTER 6 **Power grid development and its comprehensive
social and economic benefits** **197**
6.1 China's Power Grid Development in the Future 197
6.1.1 The Internal Relation Between Energy Structural
Adjustment and Power Grid Functions............................ 197
6.1.2 The Overall Situation of China's Power Flow 200
6.1.3 China's Power Grid Development in the Future 202
6.2 The Driving Effects of Power Grid on Non-fossil Energy
Development 204
6.2.1 Promote Hydropower Development and Transmission
and Efficient Utilization 204
6.2.2 Promote the Large-Scale Development and Efficient
Consumption of New Energy 206
6.2.3 Promote Safe and Economical Operation of Nuclear
Power 212
6.2.4 Comprehensive Evaluation of the Effects of Power
Grids on Promoting Non-fossil Energy
Development... 214
6.3 Comprehensive Social and Economic Benefits of Power
Grid Development..215
6.3.1 Economic Benefits... 216
6.3.2 Energy-Saving Benefits... 217
6.3.3 Resource and Environmental Benefits............................ 219

6.3.4 Safety Benefits.. 221

6.3.5 Social Benefits.. 223

CHAPTER 7   The realization routes to China's non-fossil energy development goal **225**

7.1 Overall Path 225

7.1.1 2015–20... 227

7.1.2 2020–30... 229

7.2 Specific Paths and Supporting Conditions.............................230

7.2.1 Hydropower Development Path and Supporting Conditions.. 230

7.2.2 Nuclear Power Development Path and Supporting Conditions.. 237

7.2.3 Development Path and Supporting Conditions of Nonwater Renewable Energy Sources Including Wind Power and Solar Power 239

CHAPTER 8   Policies and measures for achieving non-fossil energy development goals **241**

8.1 Follow the Strategy of "Giving Priority to Energy Conservation" and Reasonably Control Total Energy Consumption 241

8.2 Work Faster to Make a Breakthrough in Addressing Obstructions Caused by Certain Systems and Mechanisms, and Promote Non-fossil Energy Development.........................243

8.2.1 Hydropower 244

8.2.2 Nuclear Power 246

8.2.3 Wind Power 249

8.3 Improve Relevant Laws and Regulations and the Policy Support System 250

8.4 Strengthen Integrated Planning and Promote Smart Grid Construction 252

8.5 Accelerate Independent Innovation in Energy Science and Technology..253

CHAPTER 9   Main conclusions **259**

Appendix...269

Bibliography 307

Index 309

Foreword

This book is a treatise about the coordinated development of non-fossil energy and electric power. The core of the book is to answer the question "how to achieve the national goal of non-fossil energy development in a safe, economical, clean, and efficient manner." Moreover, the book, based on extensive energy and power plans and using the modeling tools such as operation simulation as support, demonstrates various scenarios of China's non-fossil energy, fossil energy, and electric power development in the upcoming 20 years. Furthermore, it analyzes several key issues such as the development, transportation, integration & consumption, and operation of non-fossil energy. In addition, it compares the cost and benefits of different scenarios and proposes the path, ideas, and suggestions for achieving the development goal of China's non-fossil energy.

The book is rich in content and has good academic and application value. It can be used as a reference book for those who study and formulate policies for the upgrade and adjustment of China's energy structure. Besides, it has important reference value for formulating energy development strategies, plans, and policies.

Preface

THE GOAL AND IMPLEMENTATION PATH OF CHINA NON-FOSSIL ENERGY DEVELOPMENT

Currently, the development of world energy and electric power is in the midst of compound transition: the energy consumption market once dominated by developed countries is turning into one that is shared by both developed and developing countries with the proportion of consumption by developing countries ascending steadily; energy supply is also undergoing fast change by which the once dominant fossil energy is gradually replaced by clean energy, typically renewable energy and nuclear energy, with their proportion still on the rise. Meanwhile, all countries around the world are scrambling for new energy technologies and hoping to occupy a strategically favorable position in the marketplace. In this transition process, the emergence and rapid development of clean energy are the core and most prominent features, which play a decisive role in forming the world energy landscape in the future. Predictably, in the upcoming several decades, the world energy structure will be pillared by petroleum, gas, coal, renewable energy, and nuclear energy.

Due to historical and resource reasons, China has been faced with a series of problems in its energy development such as irrational energy structure, severe environmental pollution, heavy carbon emission, inefficient energy use, and unbalanced regional supply and demand. Therefore, the pressure on supply security and environmental protection is huge. Under the guidance of the scientific outlook on development in recent years, China's energy development strategy has gradually become clear. First of all, we should spare no efforts to save energy, control total consumption, and enhance the utilization efficiency of energy. Secondly, we should optimize our energy structure, broaden the channels of new-energy exploration, encourage the use and proactively pursue the development of non-fossil energy, increase energy supply, and reduce environmental pollution. One may ask: "Who will supply energy to China?" The answer is clear: we need to proactively import energy from other countries but, as one of the countries with the most energy resources, China should ultimately rely upon itself to address the issue of sustainable energy development. Moreover, to achieve this outcome, China should explore and tap non-fossil energy, making it an increasingly important resource for energy development in a scientific way.

Clean and low-carbon energy such as hydro energy, nuclear energy, wind energy, biomass energy, and solar energy have huge reserves and great potential to explore in China. For some reasons, except hydro energy, various other non-fossil energy resources are basically in the primary stage of development in China. Now, despite a better research foundation and broader consensus, the level of science and technology in China is still quite below that of developed countries, and the development cost of non-fossil energy remains uncompetitive compared with that of traditional fossil energy. We should make every effort to address two important issues for the development of clean energy in China in the future. The first issue, from the perspective of technology, is how to increase the efficiency of energy utilization while reducing the development cost through proper planning, technology advancement, and technological innovation. The second issue, from the perspective of policy, is how to break the barriers of existing systems and mechanisms, and formulate fairer and more efficient policy-driven incentives to promote the development of clean energy. Since non-fossil energies such as hydro energy, nuclear energy, wind energy, and solar energy have to be transformed into electric power for use, electric power plays a key role in driving the development of non-fossil energy.

This book is a treatise about the coordinated development of non-fossil energy and electric power. It is based on years of research and coauthored by experts from State Grid Energy Research Institute Co., Ltd (SGERI). The core of the book is to answer the question of "how to achieve the national goal of non-fossil energy development in a safe, economical, clean, and efficient manner." Moreover, the book, based on extensive energy and power plans, using the modeling tools such as operation simulation, and adopting the quantitative research method in a systematic way, demonstrates various scenarios of China's non-fossil energy, fossil energy, and electric power development in the upcoming 20 years. Furthermore, the book analyzes several key issues such as the development, transportation, integration & consumption, and the cost effectiveness and price of clean energy; and proposes the path to take and policy suggestions for achieving the development goal of non-fossil energy. The book comes up with several unique creative ideas. Firstly, it includes clean energy into electric power planning, where the non-fossil energy development plan is attainable, ready to be integrated and consumed, and the development goal is quantifiable. Secondly, it adopts the method of multiscenario comparative analysis to demonstrate all the possible scenarios of non-fossil energy development in the upcoming 20 years, and the analysis results are representative, pertinent, and typical. Thirdly, based on the characteristics of non-fossil energy, it comes up with the idea that the development of power grid is of vital importance to the efficient use of clean energy. The book is well founded, highly readable, and credible as it has clear ideas and uses substantial data. The book crystallizes the in-depth thoughts and proactive exploration of workers within the electric power industry on some importation issues pertaining to energy and electric power development, particularly the exploitation and utilization of non-fossil energy. The thoughts and exploration are an important reference for the formulation of China's energy strategy and plan.

In the upcoming decades, we will still have to explore the right path for China's energy development. So far, we have not found a "Chinese pattern" that is mature, complete, and verified in practice. Based on international experience and national conditions, China will make the biggest contribution to mankind by creatively blazing a trail for sustainable economic and social development. Above all, a scientific, green, and low-carbon energy strategy will be the key factor for taking that trail.

The book not only opens up a meaningful exploration but also serves as a valuable reference for energy research. As a famous Chinese saying goes, "Reading benefits," so we believe energy workers will obtain inspirations and insights by reading this book.

Du Xiangwan

Introduction

THE GOAL AND IMPLEMENTATION PATH OF CHINA'S NON-FOSSIL ENERGY DEVELOPMENT

Energy security is an important part of national security and the cornerstone for the sustainable social and economic development in China. Meanwhile, it is also closely related with climate change and is a hot political and economic spot in the international community. Restricted by conditions of energy resources and the level of economic development, coal has always been the main energy resource in China's energy consumption structure. During the process of energy development, a series of issues such as heavy environmental pollution, huge total carbon emission, and inefficiency of energy utilization bring a tremendous pressure to the sustainable development of energy and even social and economic growth in the future. Therefore, we must upgrade and readjust China's energy supply structure. In a joint statement on climate change in Beijing in 2014, China pledged to increase the share of non-fossil energy consumption to 20% by 2030. Besides, the 18th CPC National Congress Report clearly pointed out the orientation of China's energy development. That is: our endeavor for energy development should give consideration to ecological civilization construction. We should not only pursue development but also give full consideration to environmental protection. We should make every effort to develop clean energy, as this endeavor is an indispensable way for readjusting the energy strategy, safeguarding energy security, conserving ecological environment, dealing with climate change, and fulfilling our commitment to the international community. Moreover, this endeavor is also closely related with the bigger picture of China's political, economic, and social development.

The book is written as an electric power technology publishing project of the State Grid Cooperation of China (SGCC). Based on the Scientific Outlook on Development, the project fully considers the coordinated development of traditional energy, electric power, and clean energy, grasps the new trend that the development of clean energy is ratcheted up, seeks safe, clean, economical, and efficient ways for the exploration and utilization of non-fossil energy, upgrades the configuration of energy structure, and endeavors to increase energy efficiency and guarantee energy security. Based on the Overall Optimization and Planning Model of the electric power system independently developed by the SGERI, the book constructs several complete scenarios of coordinated development of clean energy and electric power by virtue of the systematic quantitative research method, and comes up with feasible paths and policy suggestions for realizing the goal that non-fossil energy will account for 20% of the national total primary energy consumption by 2030.

The book consists of nine chapters. Chapter 1 analyzes the trend and features of world energy development, the experience in and trend of clean energy development, and the development features of electric power in the future. Chapter 2 analyzes

China's energy structure and the status quo of non-fossil energy, the situation of and challenges in China's energy development, and the thoughts about China's overall energy development in the future. Chapter 3 elaborates the general analysis method of non-fossil energy development and the overall optimization and planning method system of non-fossil energy and the electric power system. Chapter 4 analyzes the conditions and exploration potential of China's non-fossil energy resources and fossil fuel resources. Based on an analysis and prediction of the total primary energy consumption demand and the electric power in 2030, Chapter 5 makes a contrastive analysis of multiple scenarios and analyzes the development scenarios of different non-fossil energy resources from technological and economic perspectives. Chapter 6 expounds the relationship between restructuring of energy structure and development of electric power and demonstrates the core role of power grids in driving the development of non-fossil energy.

From the perspective of controlling the total amount of energy consumption and realizing the development goals of hydropower, nuclear power, wind power, and solar power generation, Chapter 7 analyzed the necessary measures and conditions needed to achieve the 20% non-fossil energy development goals.

Chapter 8 proposes possible policies and measures to promote China's non-fossil energy development in terms of system and mechanism, laws and regulations, development mode, and technological innovation. Chapter 9 summarizes the main research results.

The main research results of the book can be referenced for formulating long- and mid-term energy plans and renewable energy plans and improving China's energy development strategy. The book uses substantial data, and adopts both the qualitative research method and the quantitative research method. We hope the book will be a good reference for leaders, experts, technicians, and readers who are concerned with China's energy economy growth.

The book is coauthored by Yunzhou Zhang, Wang Yaohua, Fuqiang Zhang, Shengyu Wu, Bo Yuan, Junshu Feng, Guanjun Fu, Nan Li, and Libin Chen. Others who also made valuable contributions to the book include Bai Jianhua, Cheng Lu, Xin Songxu, Hu Bo, Zhang Qin, Zhang Dong, Jin Yanming, Liu Jun, Jia Dexiang, Chen Wei, Liang Fucui, Fu Rong, Wei Xiaoxia, Xu Gu, Wang Dun, Gao He, Li Qian, and Zhang Jinfang.

We thank former CAE vice president and academician Du Xiangwan who wrote an insightful Preface, and several other experts within the energy and electric power industry such as Zhou Xiaoqian, Ran Ying, Wang Xinmao, Ouyang Changyu, and Jiang Liping who helped a lot and gave suggestions in the writing process.

Due to the limited capabilities of the authors, the book may have errors despite our repeated efforts to improve it. We would like readers to point out the errors, if any, and correct them!

Coauthors

Review and outlook of world energy development

The exploitation and utilization of energy has marked the progress of human civilization. As a matter of fact, it is a human activity existing in all the phases of human civilization ranging from primitive civilization, agricultural civilization, industrial civilization, all the way to modern ecological civilization. While energy drives the progress of human society, human civilization also pushes forward transformation by which people exploit and use energy. Together with the growth of population and the economy as well as the technological advancement, the world energy has witnessed three major transformational stages that are intertwined and are under gradual transition. The first stage was when coal replaced firewood to become the principal energy source; the second stage was when oil took the place of coal as the leading energy resource; and the third stage was when energy structure began shifting toward plurality, which started in the second half of the 20th century.

Before the industrial revolution of the 18th century, biomass energy such as firewood was the main source of energy. After that, when the steam engine was invented and used, coal gradually replaced firewood. From the end of the 19th century to the early 20th century, with the creation of the internal combustion engine and electricity, two revolutionary technologies, the demand for fossil fuel has surged and coal became the principal energy source. After the Second World War, due to the development of the automobile industry and the emergence of multinational oil companies, oil consumption grew rapidly, resulting in the proportion of oil exceeding that of coal in energy consumption structure. Then after the discovery of a large number of natural gas fields, natural gas was quickly exploited and widely used in daily life. Since the beginning of the 21st century, oil, coal, and natural gas have become the three main sources of world energy supply, altogether accounting for more than 80% of total world energy consumption.

However, extensive use of fossil energy releases a lot of carbon dioxide, sulfur dioxide, nitrogen oxide, and smoke into the atmosphere, which in turn give rise to very severe ecological and environmental problems. To alleviate some negative effects of using fossil energy on the climate, the ecosystem, and the natural environment, the international community has put unprecedented emphasis on the exploitation and the use of non-fossil energies like hydro energy, nuclear energy, wind energy, and solar energy. The world energy industry has entered a brand-new era of low-carbon green development.

Non-Fossil Energy Development in China. https://doi.org/10.1016/B978-0-12-813106-0.00001-5

1.1 HISTORY AND STATUS QUO OF DEVELOPMENT
1.1.1 POPULATION, ECONOMY, ENVIRONMENT, AND ENERGY CONSUMPTION

The rapid increase in the world population began in modern times (since 1950). According to statistics released by the Population Division of the UN Department of Economic and Social Affairs, some 2000 years ago, the world population was about 300 million. Since 1950, the world population has been exploding. In 1999, this number exceeded 6 billion, nearly two and a half times the number in 1950. From 1965 to 1970, the global population growth rate hit a record high, reaching 2%. As shown in Fig. 1.1, the time span for every 1 billion increase in the world population has been shortened from 123 years in the 19th century to only 12 years now. After the 1970s, as the birth rate dropped in most regions around the world, the growth rate of world population began to decline. The world population surpassed 7 billion in 2011, reaching 7.35 billion in 2014.

With the rapid growth of the world's total population, the proportion of the urban population is steadily on the rise. Since 1950, the proportion of the urban population has been growing at an annual rate of about 2%; today half of the world's population is in the city.

The 20th century was one with the fastest economic growth in the course of the development of human society. The advancement of science and technology, rapid development of urbanization and industrialization, and restructuring of industry all constituted and accelerated the development of a social productive force. In the wake of that, people's production mode and life style witnessed a tremendous change. In the 20th century, the annual growth rate of the world's economy registered up to 6%.

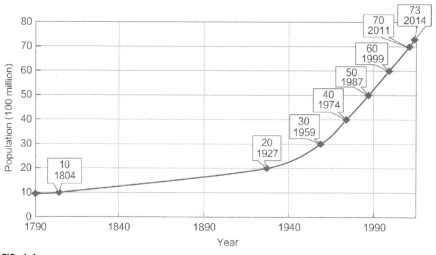

FIG. 1.1

Global population growth trend.

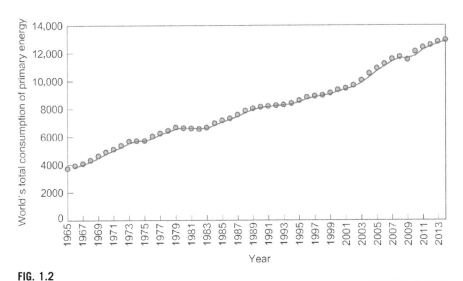

FIG. 1.2

Global total primary energy consumption growth trend.

At the beginning of the 20th century, world GDP was only 50 or 60 billion dollars. But at the end of the 20th century, this figure was as high as 30 trillion dollars. Since beginning of the 21st century, the world's economy has maintained fast growth with global GDP reaching 76.5 trillion dollars in 2014.

With ever-growing world population and economic scale, global energy consumption has continuously increased. According to statistics released by British Petroleum (BP) in June 2015, global primary energy consumption (definitions and ranges of various energy indexes are shown in Appendix I) in 1965 was equivalent to only 3.73 billion tons of standard oil, but this figure had risen to 12.93 billion tons by 2014. For the past 49 years, the annual growth rate of global energy consumption has been maintained at 20%. Global primary energy consumption from 1965 to 2014 is seen in Annexed Table A.1; its growth curve is shown in Fig. 1.2.

According to the data shown in Fig. 1.2 and Annexed Table A.1, global energy consumption grows along with population growth, economic growth, and technological progress. Meanwhile, drastic changes in energy demand coincide with major economic events or energy events. Among them, the three oil crises that occurred during the second half of the 20th century and the 2008 global financial crisis are the most influential events.

Use of fossil energy gives rise to a series of problems such as excessive emissions of greenhouse gases (GHG).[1] According to statistics released by the International

---

[1]Greenhouse gases are a generic term for gases in the atmosphere released from nature and by mankind that can absorb and reemit infrared radiation. They include water vapor, carbon dioxide, methane, nitrous oxide, ozone, Freon or chlorofluorocarbons, hydrochlorofluorocarbons, hydrofluorocarbons, perfluorocarbons, hexafluoride, and so on.

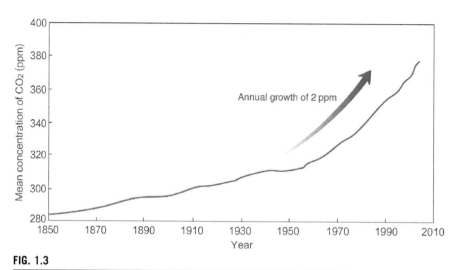

FIG. 1.3

Change in carbon dioxide concentration in the past century.

Energy Agency (IEA), of the total GHG emitted from Annex 1 countries/Parties of United Nations Framework Convention on Climate Change (UNFCCC),[2] about 80% was from carbon dioxide produced by burning fossil fuels, while for total global GHG, the proportion was 60%. The excessive burning of fossil fuels has caused a surge of carbon dioxide emissions, and the level of CO_2 in the atmosphere has increased from 270 to 400 ppm over the past 110-plus years, as shown in Fig. 1.3. It is expected that the level of CO_2 will reach 600 ppm in the middle of the 21st century. The rise rate of the average level of CO_2 in the atmosphere is accelerating: the annual rise has been 2 ppm since 2000, whereas the annual rise was only 1.5 ppm during the 1980s. Besides, excessive burning of fossil fuels also releases a huge amount of sulfur dioxide, nitrogen oxide, and smoke into the atmosphere, causing very severe ecological and environmental problems.

Since the beginning of the 21st century, climate change has become one of the most significant global environmental issues that the international community is concerned about. To protect the earth's environment and alleviate the harm caused by climate change, more than 150 countries, including China, have signed the UNFCCC since the first United Nations Conference on Environment and Development (UNCED) convened in June 1992. At the Climate Conference held in Paris at the end of 2015, about 200 contracting countries unanimously agreed to the "Paris Climate Accord" and were committed to keeping the average global temperature rise

---

[2]Annex 1 countries/Parties of United Nations Framework Convention on Climate Change (UNFCCC) includes ① the developed countries listed within the Annex 2 of UNFCCC that have donation obligations; ② countries in the midst of economic transition including the Soviet Union and countries from Northern Europe.

within or below 2°C on the basis of the preindustrial level and endeavor to maintain the temperature rise within 1.5°C. Besides, member states pledged that they will allow global GHG emission to reach the peak value as soon as possible and realize the net-zero emission in the second half of the 21st century.

1.1.2 STRUCTURE OF WORLD PRIMARY AND END-USER ENERGY CONSUMPTION STRUCTURE (EUECS)

Since the First Industrial Revolution[3] in the 1860s, fossil fuel had been in a dominating position in energy production and consumption for mankind. In 1930, coal accounted for around 70% of the total energy consumption. From then on, its proportion began to decline. Since the first oil well was exploited in 1859, oil was ahead of coal in energy consumption, jumping to the dominant position of primary energy consumption until the 1960s. In 2014, world primary energy consumption reached about 12.93 billion tons of standard oil equivalent, of which coal, oil, and natural gas, the three kinds of fossil fuels, accounted for 30.0%, 32.6%, and 23.7%, respectively.

Due to the growing concern about energy supply security and global climate change, countries around the world, one after another, endeavored to seek and develop substitute energy for the traditional fossil fuels. The relatively low-carbon fossil fuels such as natural gas and non-fossil energy were consumed so fast that they took the place of oil and coal in energy consumption. The proportion of oil to world primary energy consumption climbed to a peak in 1973 and began to drop after the first oil crisis. In 2014, this proportion dropped to 32.6%. The changing trend of world primary energy consumption structure is shown in Fig. 1.4.

Since the second half of the 20th century, the proportion of coal consumption to world primary energy consumption had continuously dropped, from 37.4% in 1965 to 25.0% in 1999. Meanwhile, the world coal consumption was also influenced by a decline in coal consumption in the developed economies and a hike in the emerging economies respectively. From the perspective of the proportion of coal's consumption trend, the proportion of the world's coal consumption had been in a rapidly decreasing phase before the 1970s; however, after that phase, driven by an economic pickup of the emerging economies, coal consumption ramped up rapidly. The drop in the proportion of coal consumption slowed down from the 1970s to 1990s when some fluctuations occurred. The turning point of the proportion of coal consumption occurred in 1999 when the major demand for coal to drive economic growth in developing countries such as China and India stimulated the rise in the proportion of coal consumption around the world. Additionally, such proportion returned to 30.9% in 2014.

---

[3]The First Industrial Revolution was a technological revolution starting from the 1760s until the middle the 18th century. It was first initiated in Britain with machines taking the place of handcraft and hand tools. The revolution began when the first working machine was created. Those steam engines were widely used as the power machine and marks the birth of the revolution.

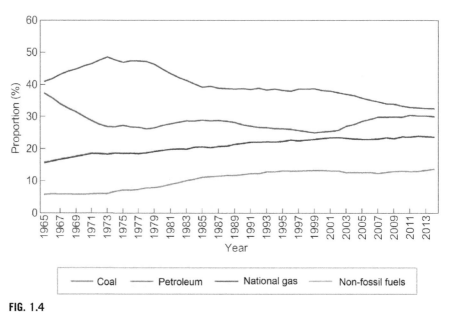

FIG. 1.4

Global total primary energy consumption mix development trend.

From 1965 to 2013, the proportion of natural gas consumption was ramped up by a 0.17 percentage point annually. In 2014, under the double influence of the world's economy slump and burgeoning of non-fossil energy, the proportion of natural gas consumption declined by 0.1 percentage point for the first time. The proportion of non-fossil energy consumption was on a steady rise from 1965 to 2014, increasing by 0.16 percentage point annually. The structure of world primary energy from 1965 to 2014 is shown in Annexed Table A.2.

The IEA released the end-user energy consumption structure (EUECS) statistics of the world, as illustrated in Table 1.1. In this structure, the proportion of fossil energy consumption continuously declined, while the proportion of electric power consumption ramped up significantly, when more and more fossil fuels like coal and natural gas were transformed into electricity. From 1971 to 2013, the proportion of fossil fuel consumption in the world EUECS such as coal, oil, and natural gas dropped by 9.1 percentage points while the proportion of electric power nearly doubled, reaching 18.0% in 2013. Features and trends showing that energy consumption is heavily relying upon electric power have become more and more evident.

1.1.3 WORLD ENERGY CONSUMPTION AND PRODUCTION BY REGIONS

1.1.3.1 World primary energy consumption by regions

From the perspective of the geographic distribution of world primary energy consumption, the Asian-Pacific region was the fastest growing zone in terms of global energy consumption, surpassing North America in 2001 and Europe in 2002, as

Table 1.1 End-user energy consumption structure (EUECS) of the world

Type of energy consumed by the end user	1971	1980	1990	2000	2005	2010	2013
Coal	14.6%	13.0%	121%	7.6%	8.4%	9.8%	11.5%
Oil	46.8%	45.3%	414%	44.2%	43.5%	41.6%	39.9%
Natural gas	142%	15.4%	152%	16.1%	15.6%	15.2%	15.1%
Electricity	8.8%	10.9%	133%	15.4%	16.4%	17.3%	18.0%
Biofuels and waste	129%	13.7%	118%	13.1%	12.8%	12.5%	12.2%
Other types	2.7%	1.7%	62%	3.6%	3.3%	3.6%	3.3%
Total (equivalent to 1 million tons of standard oil)	4256	5381	6293	7037	7878	8329	9301

Data from: IEA, Key world statistics 2015, 2015.

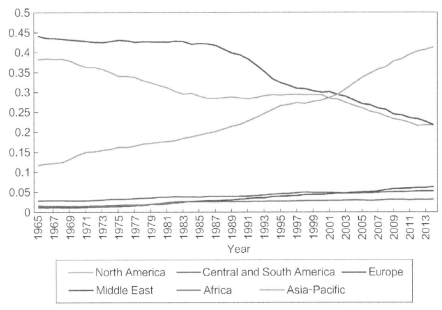

FIG. 1.5

Global total primary energy consumption geographical distributions trend.

shown in Fig. 1.5. In 2014, 41.3% of world primary energy consumption was from the Asian-Pacific region, 21.9% was from Europe, and the rest 15.0% went to Mid-Latin America, the Middle East, and Africa. The trend of energy consumption by geographic distribution was even more evident from the perspective of the newly added world primary energy consumption. Of the newly added world primary energy

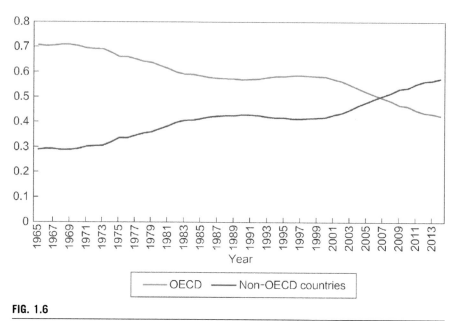

FIG. 1.6

Trend of global total primary energy consumption distribution between OECD countries and non-OECD countries.

consumption from 1965 to 2014, 53.2% went to Asian-Pacific regions while 15.2% and 12.9% went to Northern America and Europe, respectively.

From the perspective of distribution of world primary energy consumption, the energy consumption volume of non-OECD (Organization for Economic Co-operation and Development) economies, which mainly comprised developing countries, exceeded that of OECD[4] economies in 2008, as shown in Fig. 1.6. On the one hand, as the economic level of developed countries entered into a postindustrial stage, the industrial structure of which was shifting to low-energy consumption and high productivity, the manufacturing industry, with high energy consumption, was shifting to developing countries and developed countries, laid great emphasis on improving energy efficiency and was achieving substantial results. On the other hand, emerging economies were entering into the development phase of industrialization and urbanization and economic growth increased the demand for energy consumption.

---

[4]At present, OECD has 34 member states, including Australia, Austria, Belgium, Canada, the Czech Republic, Denmark, Finland, France, Germany, Greece, Hungary, Iceland, Ireland, Italy, Japan, South Korea, Luxembourg, Mexico, the Netherlands, New Zealand, Norway, Poland, Portugal, Slovakia, Spain, Sweden, Switzerland, Turkey, the United Kingdom, the United States, Israel, Slovenia, Chile, and Estonia.

The distribution of the world's total primary energy consumption from 1965 to 2014 is shown in Annexed Table A.3.

1.1.3.2 World coal consumption and production by regions

World coal consumption and production by regions in 2014 are shown in Fig. 1.7. World coal consumption and production by regions from 1965 to 2014 are shown in Annexed Table A.4 and Annexed Table A.5, respectively. According to Fig. 1.7, Annexed Table A.4, and Annexed Table A.5, world coal consumption and production by regions were basically balanced. The global trade volume of coal in 2014 only accounted for 16.9% of total world coal consumption.

1.1.3.3 World oil consumption and production by regions

World oil consumption and production by regions was evidently unbalanced, with the proportion of global trade volume of oil to world oil consumption on a constant rise. World oil consumption and production by regions in 2014 are shown in Fig. 1.8. In 2014, the global trade volume of oil accounted for 61.6% of world's total oil

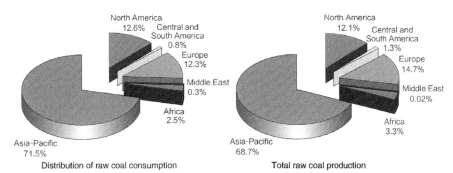

FIG. 1.7

Global raw coal consumption and production distribution in 2014.

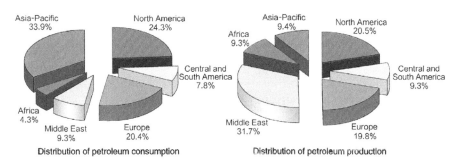

FIG. 1.8

Global petroleum consumption and production distribution in 2014.

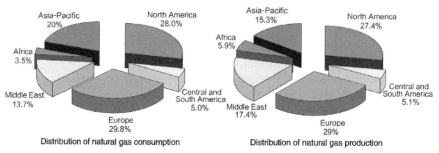

FIG. 1.9

Global natural gas consumption and production distribution in 2014.

consumption, while in 1980, this proportion was 51.4%. World oil consumption and production by regions from 1965 to 2014 are shown in Annexed Table A.6 and Annexed Table A.7, respectively.

1.1.3.4 World natural gas consumption and production by regions

World natural gas consumption and production by regions were now in balance on the whole, as shown in Fig. 1.9. However, the global natural gas trade was now rapidly ascending. In 2014, global trade volume of natural gas accounted for 29.4% of world natural gas consumption. World natural gas consumption and production by regions from 1965 to 2014 are shown in Annexed Table A.8 and Annexed Table A.9, respectively.

1.2 ENERGY DEVELOPMENT STRATEGY AND ENLIGHTENMENT FROM PRINCIPAL DEVELOPED COUNTRIES

The endeavor to guarantee energy supply security (such as the supply security of oil, natural gas, and electric power) and to alleviate the negative influence of energy production and consumption on social and ecological environment had become the main standpoints for the restructuring of world energy development strategy. The US Energy Independence Strategy, EU Energy Strategy 2020–2050, and Japan's New Energy Strategy, will together, have a significant influence on international energy development.

1.2.1 THE US ENERGY INDEPENDENCE STRATEGY

In the wake of the two oil crises in the 1970s, energy security had increasingly become the focus of some principal countries around the world. Against this backdrop, the United States came up with the Energy Independence Strategy. Then, after long-term efforts in finding the right policies and improving technology readiness, the US dependency on oil and natural gas import decreased gradually and energy security was strengthened significantly.

1.2.1.1 Basic conditions of the US Energy Independence Strategy

The core of the US Energy Independence Strategy was "Independence of Oil." It mainly referred that the United States should decrease dependency on oil import from other countries, particularly from the Middle East. The United States was the largest oil consumer in the world whose oil consumption had accounted for a quarter of world's total for a long time. Among these, 50%–60% relied upon import, most of which were from the Middle East, Canada, and Middle and Southern Latin America. During the period of two oil crises, President Nixon first called for the "Energy Independence" idea. From then on, all successive American governments inherited the idea and gradually began to engage in policy practice and technological readiness. In 2007, the Bush administration released a guideline—"*Energy Independence and Security Act*." After President Obama was sworn in, he took 'Energy Independence' as the core of his energy policy. In 2011, the US Department of Energy promulgated "*Blueprint for a Secure Energy Future*," making a commitment that by 2025, oil imports would drop by one-third, on the basis of imports in 2008, to 320 million tons.

In recent years, the US Energy Independence Strategy had made remarkable progress. The primary energy dependency on imports began to drop since 2006. In comparison, from 1982 to 2005, America's primary energy dependency on imports ramped up from 8.9% to 30.8%—the record high. But by the end of 2014, it dropped to around 26%. Along with a gradual decrease in energy dependency on imports, the proportion of oil imports to oil consumption was on a continuous decline as well. In 2014, America's oil imports dropped from 13 million barrels per day in 2005 to fewer than 4.9 million barrels per day, with the proportion of oil to primary energy consumption dropping from 60% to 37%. The remarkable progress made by America's Energy Independence Strategy in the past few years is presented in Table 1.2.

Thanks to the comprehensive measures to enhance control of energy supply, America's energy independence seemed possible. These measures included increasing domestic energy production, decentralizing energy import channels, exploiting alternative energy, and improving energy efficiency. Among these, increasing domestic energy production referred to accelerating the exploitation of shale gas and shale oil, stressing the exploitation of sea oil resources, and stepping up efforts

Table 1.2 Remarkable progress made by America's energy independence strategy in the past few years

Year	Progress made
2009	The US produced 624 billion cubic meters of natural gas, surpassing Russia to become the world's no. 1 natural gas producer
2010	US dependency on foreign oil (including refined oil products) was 49.3%, dropping to below the "warning line" of 50% for the first time since 1997
2011	The US turned from a net importer into a net exporter of product oil for the first time since 1949

to ensure a diverse electric power supply; decentralizing energy import channels referred to reducing the dependency on oil imports from the Middle East and increasing oil and gas import from countries within the Americas; exploiting alternative energy and improving energy efficiency referred to stepping up efforts on biofuel and energy-saving vehicles.

Energy Independence Strategy was an extension of America's energy security philosophy and was increasingly receiving all-round support from technological, diplomatic, financial, and military fields. In 2011, the Bureau of Energy Resource (BER), affiliated to the US Department of State, was established. As an important coordinator on America's global energy policy, it was partnered closely, by the US Department of State, the Export-Import Bank of the United States, and the Trade and Development Agency, to formulate an international energy strategy. On energy diplomacy, the BER of the US Department of State could leverage the information, revealed by sources from US embassies and partners around the world, and exert its influence. In addition, as the US Department of Energy specialized in technology, it was inclined to provide consultative advice and share this specialized knowledge. Furthermore, it was a long-term tradition for the United States to leverage its military presence and to have control over its overseas energy issues.

1.2.1.2 Conditions of shale gas exploitation in the United States

In recent years, exploitation and use of shale gas in the United States registered remarkable progress. The main factors behind this included suitable geological structure and technology advancement. First of all, the geological structure of the United States was very favorable for the accumulation and enrichment of shale gas, allowing it to have huge shale gas reserves. The unique geological structure was particularly suitable for using the fracturing technique to extract shale gas. In addition, after having grasped two important techniques (horizontal drilling and hydraulic fracturing) for shale gas exploitation in 2003, the United States achieved remarkable progress in shale gas exploitation. The shale gas production surged from 20 billion cubic meters (accounting for 3.6% of America's total natural gas production volume) in 2005 to 380 billion cubic meters (accounting for 40% of the total) in 2014, as shown in Fig. 1.10.

The fast development of shale gas production in the United States will persist and will drive a faster growth of global shale gas production. Extensive shale gas reserves and high-rising investment fever will continuously promote the rapid growth of shale gas production. The US Energy Information Administration (EIA) had an expectation that by 2040, America's shale gas production would reach 550 billon metric meters with an annual growth of 2%. Led by the United States, it was expected that global shale gas production would witness faster growth. China, Mexico, Canada, Argentina, and Poland have great potential for shale gas exploitation. All these countries will accelerate their process of grasping the techniques for shale gas exploration and exploitation and scale production of shale gas.

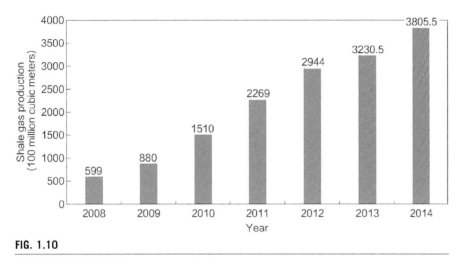

FIG. 1.10

America's shale gas production from 2008.

1.2.1.3 Several points about America's Energy Independence Strategy

America's Energy Independence Strategy aimed to step up the controllability of energy supply in terms of quantity and price and sought to constantly lower energy prices by developing energy technology, thus creating favorable conditions and increasing America's competitive power in the real economy. Against the backdrop of globalization, energy independence did not necessarily mean that the United States would become isolated in energy development. Nor did it mean the United States would stop seeking to have a tighter grip on the world's energy resources. The United States can reduce its dependency on oil imports from other countries, though total energy independence did not mean abiding by its long-holding philosophy of making overall plans for, and leveraging on, both international and domestic energy resources. Judging by America's actions toward the Middle East and the neighboring countries near the South China Sea, a tightening grip on the world energy resources will remain its primary interest.

Influenced by America's Energy Independence Strategy, the global energy balance by regions will see change on different stages: America will increase energy imports from neighboring countries while the Middle East will step up cooperation with Asia. At present, 49% of America's oil imports are from countries in the Americas; the proportion from the Middle East was reduced from 28.6% in 2001 to about 21% in 2014. It was expected that by 2020, the imports from only Canada and Mexico could meet America's oil demand. In response, regions like the Middle East, which exported oil and gas turned to other importers by stepping up cooperation with emerging economies like China and India.

Driven by the growth of America's shale gas production, global unconventional natural gas production will also see relatively rapid growth. This will continuously promote the position of natural gas in world energy structure. After breaking the bottleneck in technology, America's shale gas production registered rapid growth while China, Mexico, Canada, Argentina, and Poland were taking immediate actions to grasp the techniques in exploring and exploiting unconventional natural gas such as shale gas. Therefore, the IEA raise expectation in production and consumption of global natural gas, making natural gas the fastest growing energy among all fossil energy resources.

In the mid-term and long term, the strategic position of a clean energy generation such as wind power, solar power, and nuclear power will remain unchanged in America's energy development, with electric power as the main carrier for the development of America's clean energy generation. In the short term, negative factors such as price cuts in natural gas, due to the rapid growth of shale gas exploitation and production, and sluggish economic recovery will cast a negative impact on America's clean energy generation. But in the mid-term and long term, the clean energy generation will play a key role in coping with climate change and recovering the economy. Just as President Obama said in the State of the Union address in 2011, by 2035, it will reach the goal that 80% of electric power in the United States will be generated by clean power and the goal will remain unshakable. As more clean energy will be transformed into electric power, the position of the electric power industry is expected to be further elevated.

1.2.2 EU ENERGY STRATEGY 2020, 2030, AND 2050

In 2007, the EU set the mid-term energy and climate goal. To ensure its achievement and make a further step in building a more competitive, secure, and sustainable energy system, the EU promulgated *"Energy 2020"* in 2010, *"Energy Roadmap 2050"* in 2011, and *"2030 Climate and Energy Policy Framework"* in 2014, aiming to optimize the European energy system, create more jobs, promote economic growth across the board, and step up efforts to protect the climate and the environment by taking multiple measures simultaneously, such as saving energy and cutting emissions, increasing the proportion of renewable energy, improving energy efficiency, adopting stimulating policies, and reforming the system.

1.2.2.1 Energy development goal

The EU's energy package plan made by leaders from the 27 EU member states in March 2007 established the medium-term *"20-20-20"* goal relating to energy and climate change. Then the goal was further deepened and extended successively. The 2020, 2030, and 2050 oriented energy development goal laid great efforts on scaling up energy conservation and emission reduction, increasing the proportion of renewable energy, and improving energy efficiency. Meanwhile, it also included stimulus policies and the orientation of reforming energy system. Details can be seen in Table 1.3.

As illustrated in Table 1.4, the EU's Emission Trading Scheme (EUETS) aimed to cut emissions by 43% based on the emissions of 2005. By the end of 2020, the total emission quota of EUETS members will be down by 2.2% annually while the non-EUETS members will cut emission by 30% on the 2005 basis. By 2030, the EU will

Table 1.3 EU energy development goals for 2020, 2030, and 2050

Content	2020	2030	2050
GHG emissions[a]	Down 20%	Down 40%	80%
Proportion of renewable energy	Reaching 20%	Reaching 27% at least	–
Energy efficiency	Up 20%	Up 27% at least	–

[a]Compared with the level of year 1990.
Data from: EuroStat.

Table 1.4 EU's 2020 and 2030 GHG emission reduction targets

Item	EUETS	Non-EUETS	Total emission reduction
2020	21%	10%	20%
2030	43%	30%	40%

Data from: EuroStat.

cut emissions by 40%—a periodic achievement of the EU's 2050 goal that was expected to cut emissions by 80%. The detail is presented in Table 1.4.

1.2.2.2 Key points for strategy implementation

The policy framework of "*Energy 2020*" included five key points: energy conservation, market integration, energy security, technology innovation, and a deepened international exchange.

(1) The EU should improve energy efficiency to the maximum in the fields of construction and transport. By applying an energy management mechanism, it should increase the efficiency of industries and step up energy supply efficiency.

(2) The EU should enact and amend laws and regulations that govern the efficiency market of the EU; make construction plans for infrastructure; and formulate an appropriate financing framework.

(3) The EU should offer user-friendly energy services. By providing affordable solutions for energy consumption, it will help consumers to be more engaged in energy markets and persistently ensure energy supply security.

(4) By launching four new large-scale European projects, it can ensure the EU's technological competitiveness. Besides, it formulates a financial support plan for a cutting-edge low-carbon technology worth 1 billion Euros.

(5) It will integrate the energy market and unify the monitoring and supervision mechanism with neighboring countries and promote nuclear security and nuclear nonproliferation standards around the globe.

"*2030 Climate and Energy Policy Framework*" showed relatively significant characteristics in terms of setting goals for emission reduction, quota target of renewable energy and management, and reform.

(1) The emission reduction goal was to be achieved by EU member states and not by virtue of international reduction quotas. Members of EUETS and non-EUETS jointly accomplish this work.

(2) The EU argued that renewable energy would play an important role in transforming current EU energy systems into more competitive, secure, and sustainable new systems. Given the disparity between the member states on renewable energy development and energy resources endowment, the EU clearly proposed that the 2030 renewable energy goal would be enacted by the EU, had a legal binding, and would not split up independent goals of the member states.

Besides, policy framework will further reform EUETS, step up competitiveness, increase security measurement indicators, and set up new energy regulation systems to make investment more transparent with a higher certainty, policies more consistent, and countries more coordinated.

The key points of *"Energy Roadmap 2050"* rested upon the proposal of structural change of energy systems and the big picture of energy policy. The roadmap made a comprehensive plan on four development routes covering energy efficiency, renewable energy, nuclear energy, and carbon capture and storage.

(1) It proposed 10 big structural reform solutions for the energy system after 200 years by conceiving multiple scenarios.

(2) The policy should put focus on reforming the management method of the electric power market, integrating regional power grids with long-distance power grids, and turning to technologies like smart power grid, energy storage, and electric vehicles.

1.2.3 ADJUSTMENT OF JAPAN'S ENERGY STRATEGY

On March 11, 2011, a severe earthquake and tsunami hit Japan, causing a nuclear leak at the Fukushima nuclear power plant and crippling the electric power system. After that, nuclear power units had been shut down successively. The overall shutdown of Japan's nuclear power plants meant Japan lost its installed capacity of electric power by around 20%. To compensate the shortfall of nuclear power, Japan engaged positively in power strategy adjustments and released the *"Energy Reform Strategy"* in 2016, stressing the development of renewable energy and improved energy efficiency playing an important role in ensuring Japan's energy supply and security.

1.2.3.1 Japan's postearthquake denuclearization plan and energy strategy adjustment

Before the Fukushima earthquake disaster, Japan had 54 nuclear reactors in operation, belonging to 10 electric power companies with a total installed capacity of 48.96 GW. After the earthquake, it had only 11 nuclear reactors in operation by September 2011, with a total installed capacity of 9.86 GW.

Under the pressure from people's opposition to nuclear power, local governments required that nuclear power plants within their administration should not be plunged back into use without check, after regular shutdown inspections. Therefore, a great many nuclear power facilities were idle, causing tension surrounding the supply of electric power. The Japanese government announced that it would shut down all nuclear power plants in operation before June 2012. Confronted by a severe shortage of electric power supply during the summer, the Japanese government was forced to suspend the denuclearization plan. In May 2015, Japan's last nuclear reactor unit, the No. 3 nuclear power unit of the Tomari Nuclear Power Plant administered by the Hokkaido Electric Power Company, was shut down. In 2015, the No. 1 nuclear power unit of the Sendai Nuclear Power Plant governed by the Kyushu Electric Power Company passed inspection and became the first reactor to be restarted. Right now, only the Sendai Nuclear Power Plant successfully resumed operation within Japan, along with the No. 1 and No. 2 units of the Takahama Nuclear Power Plant, affiliated to the Kansai Electric Power Company, passing the inspection.

The electric power shortage crisis posed a new challenge for Japan after the denuclearization plan, forcing Japan to make adjustments to its energy development strategy. On the one hand, the Japanese government stepped up efforts to exploit renewable energy power generation like photovoltaic (PV) power and wind power. By technological innovation, Japan planned to increase the proportion of its installed capacity of renewable energy, such as solar energy to above 20% before 2025, installing solar panels on the rooftops of 10 million households. Based on analysis of the Ministry of the Environment of Japan, PV energy, wind power, small hydropower (below 30,000 kW), and geothermal power had great potential for electric power generation, with the maximum installed capacity reaching 72, 410, 4.3, and 5.2 GW, respectively. On the other hand, the Japanese government was attaching greater importance to improving energy efficiency. By nurturing a new philosophy on energy consumption, the government planned to improve energy efficiency in households and businesses. In the upcoming years, it would continue to develop more advanced power-generating technologies like a more efficient Integrated Gasification Combined Cycle (IGCC), and gas units for power generation. By relying upon smart power grid technology and various advanced energy-saving equipment and financing and taxation stimulation policies, the government would promote a nationwide energy and electricity conservation initiative.

1.2.3.2 Japan's new energy reform strategy

In 2016, the Ministry of Economy, Trade, and Industry of Japan promulgated the "*Energy Reform Strategy*." By increasing the proportion of renewable energy, improving energy efficiency in households, industry, and transportation, and constructing a new energy supply system, the strategy aimed to increase energy investment, in achieving goals so that the proportion of renewable energy should reach between 22% and 24% by 2030 and green house emissions being cut by 26% on the 2013 basis.

The strategy planned to increase the proportion of renewable energy by focusing on three points: reforming price policy on renewable energy generation, improving the allocation capacity of the electric power system, and optimizing management mechanism. Firstly, it would reformulate on-grid price policy on renewable energy generation, reform the certification system of on-grid price, arrange subsidy projects in an appropriate way, and try to best reduce the burden on electric power consumers while ensuring integration and consumption of renewable energy. Secondly, it would provide two different electric power supply solutions for businesses and households, with the commercial PV power as the core in the "forerunner system." Moreover, it would enhance the coordination between the power transmission and distribution (PTD) equipment and the electric power system, and thus strengthening the capacity of PTD. Thirdly, it would optimize management methods and improve efficiency, cutting time for environmental evaluation by half.

The strategy aimed to improve energy efficiency in three areas: households, industry, and transportation. It would promote wide use of energy-saving home appliances, making it standard for households to become energy-saving by 2020, by building new houses with zero energy consumption in equilibrium and modifying current households; launch the energy conservation "forerunner system" for industrial power consumption by encouraging plants to recycle waste heat and formulating energy-saving evaluation regulations; popularize and promote new energy vehicles.

The strategy planned to construct a new energy supply system from three aspects: building a new electric power market, reducing emissions from coal power plants, and promoting hydrogen power. Firstly, by building a "surplus electric power trade market" and a "virtual power plant," it would promote energy conservation activity and ensure a stable power supply for regions. Secondly, by further reducing CO_2 emission from coal power plants, it would ensure achieving the goal of maintaining CO_2 emission coefficient at $0.3\,kg/kWh$; and by promoting the wide use of hydrogen power, trying to increase the proportion of hydrogen power, and developing technology of building hydrogen supply system in a large scale, it aimed to build a hydrogen supply system without carbon.

1.3 DEVELOPMENT EXPERIENCE AND TREND OF WORLD CLEAN ENERGY

By observing a philosophy on energy strategy adjustment of developed countries and regions like the United States, the EU, and Japan, a consensus has been reached to step up efforts in developing renewable energy like wind power and solar power and so the worldwide development of wind power and solar power entered an accelerating-growth phase. The renewable energy development goals of the world's principal countries (organizations) are summarized in Table 1.5.

European countries and America are stepping up efforts to develop clean energy, and particularly where it involves integrating nonhydrorenewable energy like wind energy into the power grid, they have made good trials in terms of orderly guidance, plan and research, grid-connection management, and optimized operation, and have gained remarkable results in practice.

Table 1.5 Renewable energy development goals of world's principal countries (organizations)

Country (organization)	Main content	Time of release
EU	Established "20-20-20" medium-term energy and climate change targets: GHG emission down 20% from the level of 1990 (preferably 30% if conditions permit); the proportion of renewable energy up 20%; energy efficiency up 20% from the level of 1995	2207
	Proposed a "2030 Climate and Energy Policy Framework:" the level of carbon emissions in 2030 will be 40% lower than that of 1990, with renewable energy accounting for at least 27% of total energy consumption	2214
	Released a "2050 Energy Roadmap:" carbon emissions in 2050 will be 80%–95% less than in 1990; proportion of renewable energy to total energy demand will increase from 1% in 2010 to more than 55% in 2050	2011
Japan	By 2025, installed capacity of renewable energy including solar energy will account for >20% of the country's total installed capacity	2011
United States	At least 80% of electricity will come from clean energy by 2035	2011

Data from: United Nations Framework Convention on Climate Change.

1.3.1 DEVELOPMENT EXPERIENCE

1.3.1.1 Orderly guidance leading to the appropriate development of clean energy

Based on economic and social development and the electric power system framework, governments should give orderly guidance for the development of clean energy and identify the appropriate development orientation and trend.

Since the beginning of the 21st century, many countries in Europe have made clear goals for further developing renewable energy; and subsequently, they have issued a series of compensation measures and rewarded methods of renewable energy generation. On January 1, 2009, Germany began to endorse the *"Renewable Energy Act 2009,"* raising the compensation standard for renewable energy generation such as wind energy, solar energy, biomass energy, and geothermal energy and put great effort into the development of renewable energy. In December 2008, the Ministry of Environment in France released the renewable energy development package. The package included 50 measures covering areas in biomass energy, wind energy, geothermal energy, solar energy, and hydropower.

Since 2013, the economy in Europe has experienced an economic downturn and electric power demands have slumped due to economic stagnation. The European countries reduced fiscal subsidies for renewable energy and guided the power industry to develop in an orderly way. Germany, Italy, the Czech Republic, and Spain made the decision to cut the new energy subsidy. The policy adjustment on PV

Table 1.6 Subsidy reduction on PV power in some European countries

Country	Policy	Measures
Germany	Renewable Energy Law 2014 was amended, cutting down PV subsidies by a large margin	Strictly limiting the annual amount of newly installed capacity; canceling subsidies to self-generated self-used power; projects with installed capacity exceeding 10 MW will no longer enjoy fixed feed-in price
Czech Republic	A bill was introduced to terminate renewable energy subsidies	New renewable energy projects after December 31, 2013 will no longer enjoy preferential feed-in tariff subsidies
Spain	Reduced subsidies to feed-in tariffs of renewable energy	The investment return rate of renewable energy projects is set to 7.5%
Italy	Set the upper limit of renewable energy subsidies	Annual subsidy expenditure is limited to 500 million euros, and the upper limit of cumulative annual subsidy expenditure 6.5 billion euros
Bulgaria	Substantially cut subsidies to solar and wind power	Repeated cuts in subsidies.
Poland	Laws were amended to reduce renewable energy subsidies	Subsidies are provided only for PV power stations below 2 MW, and the distance between any two PV projects shall be more than 2 km
Romania	Reduced renewable energy subsidies	Reduce the issuance of green power certificates, and bring down the price ceiling for power plant to purchase green power certificates

Data from: PV-TECH website, http:/www.pv-tech.cn/.

industry saw the most dramatic change and the subsidy reduction would become a trend in the future. The subsidy reduction on PV power in some European countries is illustrated in Table 1.6.

1.3.1.2 Promote integrated plan of power generation and transmission

To achieve goals that the proportion of renewable energy will account for 20% by 2020, Europe scaled up renewable energy such as wind energy and solar energy in terms of power generation and power grid integration, improved utilization efficiency, and stepped up efforts on the plan and construction of cross-border power grid interconnection. The EU established the European Network of Transmission System Operators for Electricity (ENTSO-E), through legislation, and set up the plan and study system for European cross-border power grid interconnection. Through ENTSO-E, the EU carried out preparation work for a 10-year development plan study of the European cross-border power grid interconnection and European Wind Integration Study (EWIS).

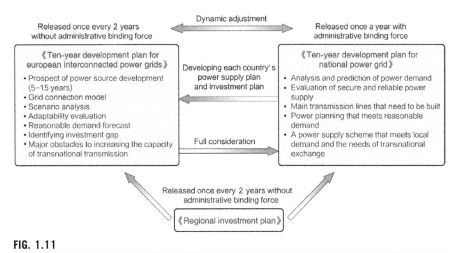

FIG. 1.11

Relationship between EU-TYNDP, Nat. TYNDP, and RIP.

According to the law package, ENTSO-E will promulgate biennially a nonlegally binding report—the *"EU-Ten Year Network Development Plan"* (EU-TYNDP), the main content of which includes the energy plan of 5–15 years, an integration model, scenario analysis, adaptive analysis of power grid, power load prediction based on rational requirement, investment gap analysis, and the analysis of the critical obstacles impeding the cross-border power transmission construction. However, the report should be coordinated with the legally binding report—"National-TYNDP" (Nat. TYNDP)—that is to be formulated annually by EU member states and should take into full consideration *"Regional Investment Plans"* (RIP). The relationship between EU-TYNDP, Nat. TYNDP, and RIP are shown in Fig. 1.11.

For instance, Spain had devised a clear wind power plan and has strictly implemented it. Since 2009, Spain had enforced the registration system for wind power predistribution. The system stipulated that any newly built wind power plant must submit a series of supporting documents including an integration consent letter from the power grid company to the administration agency of central government. Any wind power plant that was not included in the plan would not enjoy the government's subsidy on the price of electricity. In this way, Spain successfully avoided disorderly development of wind power and curbed the phenomenon of equipment lying idle, as well as waste caused by unmatched construction schedule between wind power plants and power grid.

1.3.1.3 Fully leverage flexible power regulation capacity

Foreign countries that successfully developed nonhydrorenewable energy like wind power, had a higher proportion of hydropower (including pumped storage), and oil-fired and gas-fired power units in their electric power mix, which had more flexible adjustment and more powerful capacity in peak regulation and load tracking. By the end of 2010, the installed capacity of oil-fired power, gas-fired power, and hydropower, which had excellent regulation capability, accounted for 46% of the total in Spain; compared with the figure at the end of 2000, installed capacity of wind power increased by 17.5 GW, while that of oil- and gas-fired power, which had a good regulation capability, by 18 GW. More flexible electric power structure in Spain paved the way for the integration and consumption of wind power. At 03:40 on November 9, 2010, the output proportion of wind power accounting for whole system load reached as high as 54%. Hydropower (including pumped storage) output fluctuated between −2.1 and 4.02 GW, with regulation amplitude reaching 6.12 GW; the output of oil-fired power and gas-fired power on that day fluctuated between 1.42 and 7.04 GW, with regulation amplitude reaching 5.62 GW. The regulation amplitude of all the above power units on that day reached as high as 11.74 GW, which was equivalent to 3.2 times the of wind power fluctuation on that day. The flexible regulation capability of electric power in Spain provided strong support to its integration and consumption of wind power.

As a part of the electric power installation structure in the United States, the installed capacity of single-cycle combustion gas-fired power only reached around 100 GW. Moreover, along with the slump of the price of natural gas due to the large-scale exploitation of shale gas, the installed capacity of combustion gas-fired power was quickly increasing. Due to the extreme difficulty in establishing new hydropower plants (including pumped storage), installed hydropower capacity had been basically stable, with the installed capacity of regular hydropower and pumped storage power plants being roughly maintained at 99 and 20 GW, respectively, since the early 1990s. As the potential to increase the peak load regulation capacity by establishing new regular hydropower plants and pumped storage power plants is limited, combustion gas-fired power that has a strong peak regulation capacity will be the key method for the United States to tap its renewable energy in the future.

1.3.1.4 Expand integration and consumption scope through interconnected power grid

The cross-regional and cross-border interconnection of European electricity grids further promotes large-scale integration and the consumption of clean energy such as wind power and PV power. Among the four most northern European countries, Denmark is one whose energy structure is mainly composed of thermal power and wind power. In 2014, their installed capacity of wind power was 5 GW, accounting for 30% of the total; in Norway, hydropower has a very large proportion with an installed capacity as high as 98.6%. In Sweden, energy structure is mainly composed of hydropower and nuclear power, with the former accounting for a larger proportion. Denmark and its neighboring countries have close connection in terms of electricity grids. By the end of 2015, the transmission capacity of cross-border power lines had reached 8.28 GW—that was 1.2 times the installed capacity of renewable energy. At 02:40 on December 21, 2013, the electrical load was 3.04 GW while the installed capacity of wind power reached 4.07 GW and the cross-border transmission capacity was 2.54 million, accounting for 62.4% of wind power total. The 30% proportion that the installed capacity of wind power, accounting for the total in

Denmark, was strongly underpinned by two factors: neighboring countries had sufficient and flexible electric power and feasibility of cross-border power exchange.

Clean energy developed fast in Germany, making it the country with the most installed capacity of PV power. In 2014, installed capacity of PV power was nearly 40 GW, holding 20% of the total; it had 59 cross-border transmission lines that were 220 kV or above and could transmit more than 60% of PV power to neighboring countries. In the noon of March 20, 2015, the entire Northern Hemisphere witnessed an unprecedented total solar eclipse. According to "*the evaluation report on the impact of solar eclipse 2015*" released by ENTSO-E, when the solar eclipse happened, 51% of reduced PV power of the entire European Continent was from Germany, causing a risk for overload operation of power grid. As evidence had proved, under the joint coordination of the entire European power grid, the power transmission system operators in countries like Germany maintained the power supply, balanced demand, and successfully resolved the crisis during the solar eclipse period by regulating the generated output of spare generator units.

1.3.2 DEVELOPMENT TREND

1.3.2.1 Construction plan for the big European-American power grid integration

Power grid integration will produce benefits in terms of scale. Different power grids complement each other in terms of electricity load and energy structure. Big integration can meet the demand for increasing the proportion of wind power, further promoting power generation, integration, and consumption of a renewable energy like wind power. At present, countries all around the world are accelerating the power grid integration process and increasing the integration scale; countries in different continents are establishing cross-border interconnected power grid and the global power grids show a remarkable development trend for interconnection.

1.3.2.1.1 The expansion plan for the interconnection of American power grids

The status quo and current situations of power grids: the North American Interconnected Power Grid consists of eastern power grid, western power grid, Texan power grid, and Quebec power grid in Canada—four synchronous power grids, as shown in Fig. 1.12. The total installed capacity of the North American Interconnected Power Grid is 1.1 TW, offering service to 334 million people. Besides, the electric power demand is 830 GW and the 230 V and above transmission line is about 340,000 km in length.

The United States viewed the renewable energy development as an important strategic choice. President Obama remarked in The State of the Union address in 2011 that the United States is about to realize the goal that 80% of electric power is from clean energy, wind power and PV power will witness fast development by 2030, and the installed capacity of wind power in the United States is expected to reach 300 GW by 2030. American wind power resources are mainly located in the

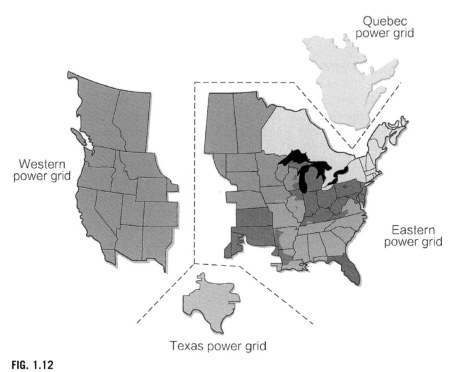

FIG. 1.12

Diagram of the North American Interconnected Power Grid.

eastern region, middle-eastern region, and Texas in the southern region, while solar power resources are mainly located in the middle and eastern regions. The first and foremost priority for the future expansion plan of American power grid is how to cope with the issue of the large-scale integration and consumption of renewable energy like wind power.

In 2003, the US Ministry of Energy issued the *"Grid 2030"* plan, clearly proposing future development vision for American power grid, in which the electricity transmission lines with higher capacity will connect eastern and western coasts of the United States, neighboring countries like Canada and Mexico, as shown in Fig. 1.13; establishment of the great grid will integrate and configure loads nationwide that are different, in terms of season and climate, thus improving the system operation efficiency. In early 2016, the Tres Amigas super power station plan was officially initiated. Tres Amigas, located in New Mexico, will make it possible to interconnect the eastern power grid, western power grid, and Texas power grid in North America, and will preliminarily achieve the goal of transmitting wind power of gigawatt level from the middle plain of America to eastern and western coasts, from Texas to Massachusetts, Mississippi, to Montana, and greatly improving the integration and consumption of renewable energy and the stability of the power grid.

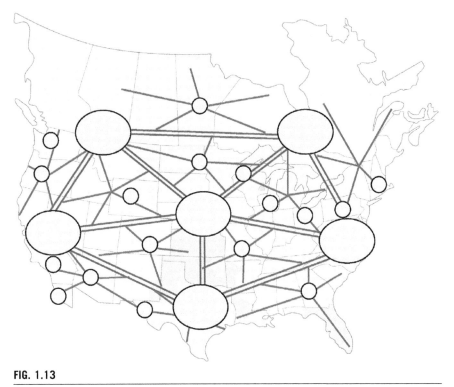

FIG. 1.13

American power grid in "Grid 2030" plan.

1.3.2.1.2 Expansion plan for European power grid integration

The status quo and current situation of the European power grid: the European Network of Transmission System Operators for Electricity (ENTSO-E) is composed of five synchronous power grid regions including the European Continent, Northern Europe, the Baltic Sea, the United Kingdom, Ireland, and two independent power grid systems including Iceland and Cyprus, as shown in Fig. 1.14. By the end of 2014, 220 kV and above transmission lines were 312.7 thousand kilometers in length, with a total installed capacity of the power grid reaching around 1.024 TW, power-generating capacity reaching 3.31 trillion kWh, electric power consumption reaching around 3.21 trillion kWh, providing service to 532 million people. The exchanged electric power among member states was about 423.6 TWh, accounting for 13% of the total electric power consumption.

The EU promised that by 2020, renewable energy would account for about 20% of the overall energy supply, with the expectancy that the power-generating capacity of renewable will reach 933 TWh, accounting for 25.5% of the power supply. The newly added electric power generated from renewable energy is mainly from wind power generated in the North Sea, the Baltic Sea, and the middle and southern areas

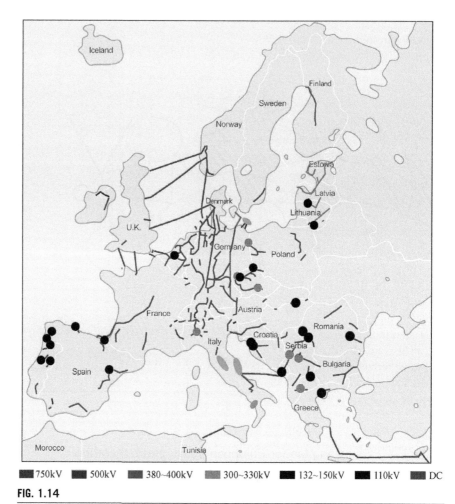

750kV   500kV   380~400kV   300~330kV   132~150kV   110kV   DC

FIG. 1.14

Diagram of ENTSO-E.

of the European continent, and from the solar power generated at Iberian Peninsula, the middle and southern areas of the European continent, and the East and South Mediterranean.

To cope with the integration and consumption of newly added renewable energy, the development goal of ENTSO-E is to integrate the large-scale centralized or decentralized wind power and solar power into the European power grid before 2020, achieving the goal of a complementary and unified operation of wind power, solar power, and hydropower, and taking full advantage of peak regulation capacity of hydropower in Sweden and Norway to improve the integration and consumption of wind power more efficiently and improve the secure and stable operation of the power grid through a high access proportion of renewable energy.

With further development of renewable energy such as wind power and solar power, and the deeper integration of European power and the electricity market, the main grid structure of the European power grid will be greatly extended and strengthened. In January 2010, neighboring countries along the North Sea, including Germany, France, Belgium, Holland, Luxemburg, Denmark, Sweden, Ireland, and the United Kingdom, officially issued the North Sea Super Power Grid Plan, proposing to connect the offshore wind turbine in Scotland, solar power array in Germany, wave power plants in Belgium and Denmark, and hydropower plants in Norway, thus forming the interconnected power grid that runs from the North Sea to northern area of the European continent. The key points of short- and middle-term plans rest upon strengthening the construction of the big cross-border and cross-region power grids of the European integration while in the long term, with the further progress made over the giant PV project in the Sahara, Africa, the European power grid will shift from the great European power grid to the European super great power grid.

1.3.2.2 Trend of development mode for wind power and solar power

1.3.2.2.1 Distributed mode will play a key assistant role

The position of renewable energy in the future energy supply system mainly rests upon its supply potential, economy, and response degree to the demand. Taking into consideration these three factors and the domestic and international research results, it can be concluded that in the global energy supply structure, distributed energy will be an important supplement to large clean energy farms, playing an assistant role in energy supply structure.

The economy of distributed generation will step up gradually. According to PV power generation roadmap (PV roadmap) report released by the IEA, it is expected that by 2050, the cost of distributed PV power generation will be between 0.07 and 0.09 dollar per kWh, while the cost of centralized PV power generation will be 0.05–0.065 dollar per kWh, as shown in Fig. 1.15.

1.3.2.2.2 Concentrating solar power is on the rise

In recent years, the global installed capacity of concentrating solar power (CSP) rapidly increased. By the end of 2009, the installed capacity of CSP was only 700 MW globally, while by the end of 2015, it had reached 4940 MW. The established CSP projects were mainly located in the United States and Spain while places like United Arab Emirates, India, Iran, Italy, Germany, and Australia, also saw some CSP projects. The installed capacity of CSP under construction and to be delivered in 2 years was about 300 MW, mainly distributed in regions like South Africa, India, the Middle East, and Mexico.

In 2016, the European Solar Thermal Electricity Association (ESTELA), Greenpeace International, and the Solar Power and Chemical Energy Systems (Solar-PACES) jointly issued the "*Solar Thermal Electricity Global Outlook 2016*" report. The report made a research on three development scenarios of CSP, as shown in Fig. 1.16. In an accelerated development scenario, it is expected that global CSP

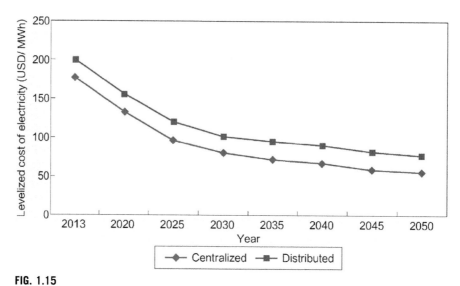

FIG. 1.15

Outlook of photovoltaic power generation cost.

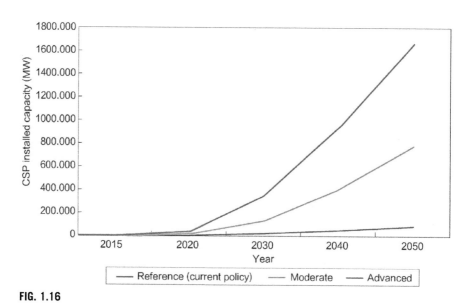

FIG. 1.16

CSP development scenarios.

installed capacity will exceed 42 GW by 2020 and 350 GW by 2030, when PV power will meet 6% of global electricity demand.

1.3.2.2.3 Offshore wind power in Europe makes fast progress

According to predictions by the European Wind Energy Council (EWEC), wind power will provide more than 50% of electricity to Europe by 2050, with a total installed capacity reaching 735 GW, of which offshore wind power (460 GW) will surpass onshore wind power (275 GW). The short- and mid-term development goal is to extensively exploit offshore wind power on a large scale in Europe. In 2015, offshore wind power in Europe witnessed fast development with the annual newly added offshore wind power integration capacity reaching 3.02 GW. The year 2015 became the fastest growing year on a year-on-year basis, growing by 108%. By the end of 2015, the installed capacity of offshore wind power in European countries was as high as 110.27 TW. The operating and approved installed capacity of offshore wind power in Europe from 2016 to 2026 will reach 37.42 GW, as illustrated in Table 1.7.

1.3.2.3 R&D and large-scale application trend of energy storage technology

Energy storage is also a key technology for promoting development of renewable energy such as wind power. The technologies for storing electricity can be generally divided into three categories: physical energy storage (pumped storage, air compression storage, freewheel energy storage), electrochemistry energy storage (sodium-sulfur cell, flow cell, lead-acid cell, lithium ion cell, nickel-cadmium cell, supercapacitor), and superconductivity electromagnetism storage. Among these, pumped storage is the most sophisticated technology while electrochemistry energy storage is making the most progress, of which sodium-sulfur cell, flow cell and lithium ion

Table 1.7 Development of offshore wind power in European countries

Capacity put into operation as of the end of 2015			Capacity approved for installation in 2016–26		
Country	Capacity (10,000 kW)	Proportion (%)	Country	Capacity (10,000 kW)	Proportion (%)
Britain	506. 1	45.9	Britain	1452. 0	55.0
Germany	329.5	29.9	Germany	691.7	26.2
Denmark	127. 1	11.5	Sweden	198.0	7.5
Belgium	71.2	6.5	Belgium	110.9	4.2
Netherlands	42.7	3.9	Ireland	100.3	3.8
Sweden	20.2	1.8	Denmark	52.8	2.0
Other countries	5.9	0.5	Other countries	34.3	1.3

Data from: Wind Europe.

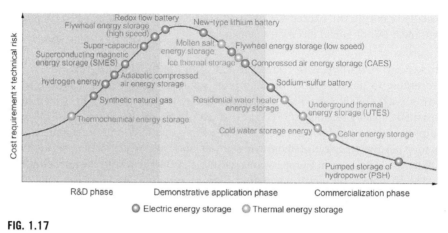

FIG. 1.17

Energy storage technology readiness level.

cell have made important breakthrough in terms of safety, energy transformation efficiency, and economy, and are gradually shifting to an industrialized application phase. At present, technology readiness level of various energy storage technologies is shown in Fig. 1.17.

At present, the United States and Japan dominate in energy storage technology. Since the 1990s, Japan has invested a huge amount of capital in large-scale research and development of energy storage technology, including providing financial support to the all-vanadium redox flow battery and sodium-sulfur cell technology. Particularly with sodium-sulfur cell, Japan has provided financial support to research and development at an early stage for free, given aid to many demonstrative projects, and continued to give subsidies after carrying out commercialized operations. Japan is in a monopoly position in terms of sodium-sulfur cell technology. After the financial crisis, the American government established large-scale energy storage technology as a key supporting technology for rejuvenating the economy and realizing the New Deal in Energy. According to the *"American Recovery and Reinvestment Act,"* the American government had appropriated 2 billion dollars to support research and development in electrochemistry energy storage cell technology for the first half of 2012. In the supportive plan for the smart power grid formulated by the US Department of Energy, the number of energy storage projects is 19, far exceeding that of other projects all together and receiving the most financial support. On August 5, 2009, President Obama announced the development plan for energy storage cells and electric cars with a total investment of up to 2.4 billion dollars. The plan covered 48 projects ranging from materials, cells, and electrically driven elements to the finished cars and charging station, bringing electric cars into the commercialized application phase. Taking into consideration the judgment of agencies like the International Renewable Energy Agency (IRENA) and the IEA, the future development roadmap for energy storage technology is shown in Fig. 1.18. The pumped storage power station will remain the main energy storage technology for the power grid

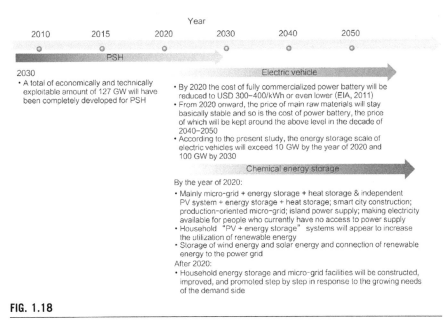

Year

| 2010 | 2015 | 2020 | 2030 | 2040 | 2050 |

PSH

2030
• A total of economically and technically exploitable amount of 127 GW will have been completely developed for PSH

Electric vehicle
• By 2020 the cost of fully commercialized power battery will be reduced to USD 300–400/kWh or even lower (EIA, 2011)
• From 2020 onward, the price of main raw materials will stay basically stable and so is the cost of power battery, the price of which will be kept around the above level in the decade of 2040–2050
• According to the present study, the energy storage scale of electric vehicles will exceed 10 GW by the year of 2020 and 100 GW by 2030

Chemical energy storage
By the year of 2020:
• Mainly micro-grid + energy storage + heat storage & independent PV system + energy storage + heat storage; smart city construction; production-oriented micro-grid; island power supply; making electricity available for people who currently have no access to power supply
• Household "PV + energy storage" systems will appear to increase the utilization of renewable energy
• Storage of wind energy and solar energy and connection of renewable energy to the power grid
After 2020:
• Household energy storage and micro-grid facilities will be constructed, improved, and promoted step by step in response to the growing needs of the demand side

FIG. 1.18

Energy storage technology development roadmap.

in 2020; lithium ion cell and flow cell are expected to be the mainstream cell technology with the best large-scale commercial prospect from 2025 to 2030, accounting for more than 50% of the global energy storage cell capacity; after 2030, the development of pumped storage will enter the period of saturation.

1.4 FUTURE DEVELOPMENT FEATURES OF WORLD ENERGY AND POWER SYSTEM

1.4.1 FEATURES OF ENERGY DEVELOPMENT

The development of non-fossil energy represented by nonhydrorenewable energy like wind power and solar power and the large-scale exploitation and utilization of unconventional oil and gas resources will become the main features of international energy development. Based on these features, world energy structure, distribution of production, and regional balance for energy resources will show their own new or strengthened development features.

Global energy structure is increasingly becoming more diversified and low-carbon. In the future, the proportion of natural gas will go up steadily. Moreover, the proportion of non-fossil energy will rise rapidly due to the fast development of renewable energy. According to the statistics released by BP, by the end of 2014, of the world primary energy consumption, oil held 32.6%, natural gas accounted for 23.7%, coal occupied 30.0%, and non-fossil energy grabbed 13.7%. It is expected that by 2035, oil, coal, and natural gas will hold a similar proportion

Table 1.8 Distribution of oil growth between 2010 and 2030 unit: 10,000 barrels/day

Country (organization)	Production increase	Country (organization)	Production increase
OPEC	800	Canadian Oil Sands	220
Non-OPEC countries of the Americas	>620	Brazil deep-water oil	200
Where American shale oil	200		

Data from: BP, World Energy Development Outlook for 2030.

from 26% to 28%. Coal consumption will reach its peak in 2020 and then go down slowly; the proportion of non-fossil energy will go up to around 38%, of which the proportion of nonhydrorenewable energy (including wind power, solar power, and biomass energy) will rise from 3.0% in 2015 to 8.0%, while the proportion of hydro-power and nuclear power will remain from 6% to 7%.[5]

Based on the energy demand from developing countries in Asia, with China as the biggest demand side, 96% growth of global primary energy consumption will be from non-OECD economies. Although the production of world oil and natural gas will converge on countries of Organization of Oil Exporting Countries (OPEC),[6] the role of some non-OPEC countries on the American continent will become more important. The oil growth distribution from 2010 to 2030 is presented in Table 1.8. With China and India as the main consumers, growth of global energy consumption will mainly come from non-OECD economies. In 2014, OPEC countries held 41%[7] of the total world oil production. In the future, the proportion will continue go up and account for 50% of the world's total. In 2014, oil production in the United State witnessed the most significant growth, producing 1.5 million barrels per day, and shale gas production is expected to increase by 2% on a yearly basis until 2040. In the next 20 years, global coal production will primarily come from China and India.

1.4.2 DEVELOPMENT TREND OF ELECTRIC POWER

In the process of ensuring a more secure energy supply and a cleaner energy structure adjustment, the power industry is no doubt in the core position. World power industry will register change in terms of power-generating energy structure and development of electricity and power grid, adapting to and promoting the low-carbon green development of world energy.

---

[5] BP, World Energy Outlook 2030, 2012.
[6] At present, OPEC has 11 member states including Saudi Arabia, Iraq, Iran, Kuwait, United Arab Emirates, Qatar, Libya, Nigeria, Algeria, and Venezuela.
[7] BP, Statistical Review of World Energy 2016.

In the future, coal will remain the primary power-generating energy around the globe. The proportion of natural gas to power-generating energy will witness a steady rise while the proportion of non-fossil energy will increase rapidly. It is expected that from 2015 to 2035, global power generation energy consumption will increase by 2.1% annually. Before 2035, coal will remain the main power-generating energy, accounting for 39% of the total. The combined proportion of nuclear power, hydropower, and other renewable energy-generated power will surpass coal in the power-generating energy structure in 2030, reaching around 40%. The proportion of natural gas will be relatively stable and remain around 20% before 2030. The world power-generating energy structure from 1990 to 2030 is shown in Fig. 1.19.

The development mode of global wind power and solar power are shifting from a small-scale distributed mode to a large-scale centralized mode, and from connecting to a power distribution network for nearby integration and consumption to connecting into a power transmission network for more balanced integration and consumption. To achieve the goal of increasing the proportion of large-scale wind power and solar power, and given the factors that renewable energy resources that are suitable for distributed development are relatively limited, and that the extent to development is rapidly increasing, the new development mode for world renewable energy in the future will show features that distributed modes and centralized modes coexist and the proportion of the latter will continue increase. The land wind power fields in Europe adopted a small-scale distributed development mode by connecting to the public power distribution network. But considering constraints such as land and

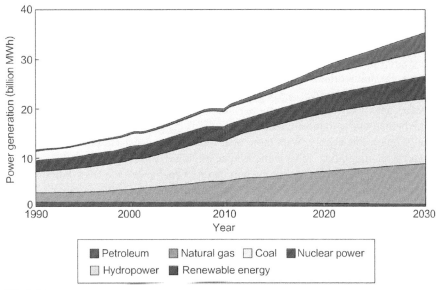

FIG. 1.19

Global power-generating mix from 1990 to 2030.

resources, the potential of the distributed development mode is limited. So, to realize the goal of increasing clean energy and carbon reduction in Europe, countries like Britain and Denmark have accelerated the exploitation of offshore wind power resources like the North Sea. Moreover, Britain has put into operation the world's largest offshore wind farm at Thanet with an installed capacity of 300 MW. The power produced in the fields will converge and connect to the European power grid for integration and consumption through high-voltage or ultrahigh-voltage transmission network. It is a trend to establish large-scale solar power stations. For this, China and the United States are leading the way.

To provide support for the fast development of wind power and solar power and to ensure secure and stable operation of the power grid, power structure in the future must become more flexible; development of power for peak regulation and application of new energy storage technology should be further emphasized. In European countries and in the United States, the huge amount of gas-fired power and hydropower that has good performance on peak regulation and is more flexible is a guarantee of the promotion of the fast development of wind power and the ensuring of a secure and stable operation for the power grid. From 2008 to 2010, the average curtailment rate of wind power was only 5%[8] in the United States. To ensure the integration and operation of an even higher proportion of wind power and solar power in the future, countries like Portugal came up with construction plan for pumped storage power stations and Europe has explored ways to increase the hydropower regulation capacity in Norway, to clip the power fluctuations of North Sea wind power and land wind power in the entire European power grid. In addition, the new energy storage devices at a specified future date will make an overall breakthrough in terms of technology and economy, thus creating convenient conditions for the large-scale utilization, and the sustainable development of nonhydrorenewable energy like wind power.

The remarkable feature of power grid development in the future rests upon the integration of a bigger and smarter power grid. To improve the use efficiency of large-scale nonhydrorenewable energy with a high proportion like wind power, it must extend the balance scope of the system power. The demand for a power transmission network with an even higher voltage class is predictable and the driving force behind this is to meet the demand for transmission, integration, and consumption of renewable energy. With more operation experience accumulated and further research on integration, most workers specializing in power grid operations around the world believe that it requires even more cross-region or even cross-border transmission capacity[9] to integrate and consume large-scale renewable energy-generated power and achieve a clean energy development goal. Reviewing the history of the

---

[8]Keven Porter, Jennifer Rogers, Ryan Wiser, Update on Wind Curtailment in Europe and North America, Consultants to the Center for Resource Solutions, June 16, 2011.

[9]The information is sourced from the US Energy Efficiency and Renewable Energy (EERE) office. The office carried out a research work on the integration of renewable energy in 2011 to fully understand the technical and management requirements for the integration of nonhydrorenewable energy like wind power. The research subjects include 33 power grid workers from 18 countries. The installation capacity of wind power they managed held 72% of world total.

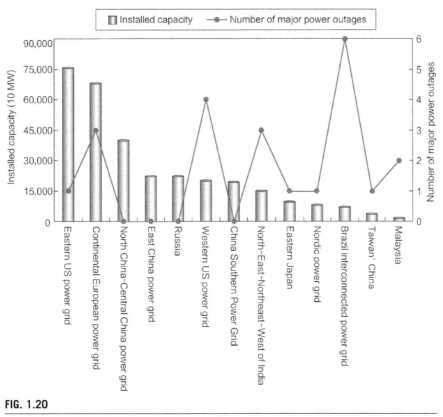

FIG. 1.20

Frequencies of major power failure of main large-scale synchronous power grids around the world does not necessarily connect with the scale of the power grid.

world power grid, it shows relatively clear stage characteristics: the power grid has gradually shifted from the phase being driven by traditional power demands, such as hydropower and coal-fire power, to the phase being jointly driven by traditional power and renewable energy-generated power, such as wind power and solar power, and in the middle and long term, being driven mainly by renewable energy-generated power like wind power and solar power. To fully tap the technological and economic advantages of AC and DC transmission technologies, respectively, and to build an even bigger synchronous power grid does not necessarily mean to increase the security risk of the power grid. As shown in Fig. 1.20, the frequency of a major power failure (lost load exceeding over 5 GW) occurring in the main large-scale synchronous power grids around the world does not necessarily connect with the scale of the power grid. As a matter of fact, a good power grid structure and a unified regulation and management are the key to ensure the secure and stable operation of the power grid.

Right now, Europe and the United States are actively making plans for expanding the power transmission network to meet the demand for the integration and consumption of wind power and solar power that hold a high proportion. The research made by an official organization of the United States came up with technical requirement for strengthening the ultrahigh-voltage AC and DC power transmission network (765 kV AC line and ±800 kV DC line) throughout the United States. To meet the challenge brought by large-scale development and the high proportion of wind power, the development of electric power still has to cope with demands from issues like power grid secure and stable control, coordinated operation of sources and power load, diversified needs from clients, resource saving, and asset utilization. It is imperative to enhance the smart level of all links such as power generation, power transmission, power transformation, power distribution, power consumption, and power scheduling, making the rise of a smart power grid a key feature of world electric power development. In addition, smart micro-grid technology[10] formed by multiple energy powers like distributed PV power, wind power, gas-fired power, and energy storage devices, is increasingly making progress.

---

[10]The definitions of micro-grid are different internationally. The definition by the American Consortium for Electric Reliability Technology Solutions (CERTS) is: micro-grid is a power supply system composed of electrical load and micro power supplies, can realize combined cooling, heating, and power; it is regulated by power-electronic equipment; viewing from the big power grid side, it is an independent controlled unit that can meet clients' demands for power quality and security. The European Commission Project Micro-grid (ECPM) defines it as: using primary energy; using micro power supplies (including uncontrolled, partially controlled, and total controlled power supplies) and capable of combined cooling, heating, and power; equipped with energy storage devices; using power electronic devices to regulate energy. The definition by the Institute of Electrical Engineering, Chinese Academy of Science is: the power generation system of micro-grid can be divided into micro gas turbines, internal combustion engines, fuel cells, solar cells, wind-driven generators, biomass energy generators, and so forth; the system installation capacity is 20–10 mW; the distributed voltage classes for clients within the power network are 380 V and 10.5 kV; if having power exchange with a public power supply system, the voltage class is based upon the real situation.

China's current situation of energy development and thinking on future development

2

China's energy endowment and current economic development stage determine that its primary-energy consumption structure (PECS), dominated by coal, is difficult to fundamentally change in the medium to long term. Overexploitation and extensive utilization of coal have brought about a series of problems such as serious environmental pollution, increased greenhouse gas emissions, and low-energy efficiency. To achieve sustainable energy development we must, on the one hand, realize the efficient and clean utilization of coal and other fossil energies, and on the other hand, continuously adjust and optimize our energy consumption structures by vigorously developing clean, low-carbon, and high-efficiency non-fossil energies.

2.1 CURRENT ENERGY STRUCTURE AND NON-FOSSIL ENERGY UTILIZATION IN CHINA

Compared with the primary energy structure, power-generating energy structure, and final energy consumption structure of other major energy-consuming countries, the proportion of total energy consumption of high-quality fossil and non-fossil energies such as oil and natural gas is low in China. This is determined by the fact that the country abounds in coal but lacks oil and natural gas and that the development and utilization scale of hydraulic, nuclear, wind, and solar power remains small in comparison with that of conventional energy.

2.1.1 CURRENT PRIMARY ENERGY STRUCTURE AND NON-FOSSIL ENERGY UTILIZATION

Over the years of reform and opening up there has been an evolution in China's energy structure where the proportion of coal shows a trend of decline. Fig. 2.1 illustrates the evolution of the country's PECS between 1980 and 2015. In the period 1997–2001 the tension between energy demand and supply was eased to some extent: the proportion of coal fell from 73.5% in 1996 to 68.5% in 2002, a drop of 5 percentage points; the proportion of oil rose from 18.7% to 21%, up 2.3 percentage points; that of natural gas increased from 1.8% to 2.3%, an increment of 0.5

Non-Fossil Energy Development in China. https://doi.org/10.1016/B978-0-12-813106-0.00002-7

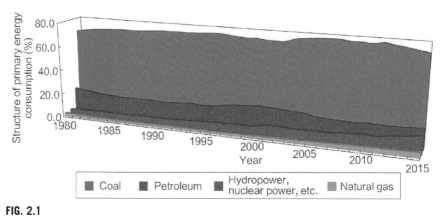

FIG. 2.1

Evolution of China's PECS in the period 1980–2015.

Data from: China Statistical Yearbook 2016.

percentage point; and that of hydraulic, nuclear, wind and other non-fossil energies climbed from 6.0% to 8.2%, an increase of 2.2 percentage points.

The Chinese economy started a new round of rapid development in 2000 featured by large-scale construction of heavy industries and a rapidly growing energy demand. The growth of energy demand slowed down in 2008 due to the impact of the global financial crisis. Later, the world saw a wave of clean energy development, with major energy-consuming countries beginning to tap clean energy as a measure to cope with the financial crisis and ensure energy security. In this context, China achieved rapid growth in clean energy power generation: during the period 2009–15, the proportion of coal in its primary energy structure fell from 71.6% to 64.0%, down 7.6 percentage points, while that of hydraulic, nuclear, wind, and other non-fossil energies in total power generation rose from 8.5% to 12.0%, up 3.5 percentage points.

In 2015 China's total primary energy consumption was equivalent to 4.3 billion tons of standard coal, where coal consumption was 2.75 billion tons, accounting for 64%; oil consumption was 7.8 billion tons, accounting for 18.1%; natural gas 193.2 billion cubic meters, 5.9%; hydraulic, nuclear, wind, and solar power generation was 1506.8 TWh, accounting for 12.0%.

Based on BP energy statistics, Table 2.1 presents the PECS of the world's major energy-consuming countries in 2014. The PECS of the United States, France, Germany, and South Korea was dominated by oil, which accounts for more than 30% of their PECS, followed by coal (except for France), and next by natural gas which accounts for about 15% (except for France and the United States). In general, the world's PECS is dominated by fossil energy while the share of non-fossil energy is less than 20%. Due to a high proportion of nuclear power in its PECS, France is the world's only major energy-consuming country whose proportion of clean energy in its PECS reaches 50%.

Table 2.1 The PECS of major energy-consuming countries in the world in 2014 (%)

Type of energy	The world	United States	Japan	South Korea	Germany	France	China
Coal	30.03	19.72	27.74	31.04	24.89	3.79	66.03
Oil	32.57	36.37	43.16	39.54	35.85	32.38	17.51
Natural gas	23.71	30.25	22.20	15.74	20.51	13.60	5.62
Nuclear energy	4.44	8.26	0.00	12.96	7.06	41.52	0.96
Hydropower	6.80	2.57	4.34	0.31	1.49	5.97	8.10
Wind energy	1.24	1.81	0.25	0.11	4.07	1.54	1.21
Solar energy	0.33	0.18	0.96	0.20	2.54	0.56	0.22
Other types of clean energy	0.89	0.84	1.34	0.11	3.58	0.64	0.36
Total share of clean energy	13.69	13.66	6.89	13.68	18.75	50.24	10.85

Note: *The data of China comes from BP Because BP and the National Bureau of Statistics of China adopt different methods when converting hydropower and nuclear power into standard quantities, the PECS calculated using the former's data is significantly different from that calculated using the latter's. BP adopts the thermal equivalent method while China adopts the coal consumption method. Other data comes from BP-released Statistical Review of World Energy 2015.*

Although the PECS of China, calculated using the data released by BP, is different from that calculated using the data released by the National Bureau of Statistics of China due to different conversion methods, both calculation results show that the proportion of fossil energy in China's PECS is too high, while that of clean energy is too low. Specifically, the proportion of coal in China is at least 35 percentage points higher than that in developed countries while the shares of oil, natural gas, and nuclear power are, respectively, 15, 10, and 10 percentage points lower than those in developed countries.

On the basis of PECS we can get China's primary energy production structure (PEPS) by taking import, export, stock change, recovery of energy, and so on into comprehensive consideration. China's dependence on import oil has been above 50% in recent years, so the gap between oil production and oil consumption is the main reason for the gap between PEPS and PECS. In 2015, the total primary energy production in China was equivalent to 3.62 billion tons of standard coal, with coal, oil, natural gas, and non-fossil energy such as hydropower and nuclear power accounting for 72.1%, 8.5%, 4.9%, and 14.5%, respectively; the proportion of oil in PEPS was 9.6 percentage points lower than that in PECS while the share of coal in PEPS was 8.1 percentage points higher than that in PECS. On the whole, the characteristics of PEPS are similar to those of PECS in China: high-quality fossil energy is insufficient and the proportion of clean energy is low.

2.1.2 STRUCTURE OF POWER-GENERATING ENERGY AND UTILIZATION OF NON-FOSSIL ENERGY

In 2015 China's installed capacities for nuclear power, hydropower (including pumped-storage power stations), wind power, solar power, and biomass power were, respectively, 27.17, 319.53, 1.1, 42.63, and 10.30 GW, each of which generated 171.4, 1111.7, 185.3, 38.5, and 52 TWh of electricity, respectively.

Table 2.2 illustrates the power generation structure of China in the period 2000–15. As seen from the table, there is no significant change in the overall structure. The proportion of thermal power in total power generation fell slightly from 82.1% in 2000 to 73.6% in 2015, down 8.5 percentage points; the share of hydropower increased by a small margin from 16.4% to 19.5% in the same period. China's first nuclear power unit was put into operation in 1993, and the proportion of nuclear power in the total power generation was 3.0% in 2015. Wind, solar, and biomass power started from scratch, accounting for about 3.9% of total power generation in 2015. Thanks to the vigorous development of wind and solar power generation in recent years, the proportion of clean energy power generation has been increased.

The power generation structure of China is significantly different from that of some developed countries. In the latter's power generation structure, nuclear power and gas-fired power take up a big proportion while coal-fired power accounts for less than 50%; the former's case is just the opposite: a high proportion of coal-fired power but a low proportion of nuclear and gas-fired power. According to a calculation based

Table 2.2 Power generation structure of China %

Category	2000	2005	2007	2008	2010	2011	2012	2013	2014	2015
Thermal power in total	82.1	81.8	83.1	80.8	80.8	82.4	78.7	78.6	75.4	73.6
Coal-fired power	78.3	78.9	80.9	78.9	78.5	78.2	74.4	74	70.4	67.7
Oil-fired power	3.4	2.4	1.0	0.7	–	–	0.1	0.1	0.1	–
Gas-fired power	0.4	0.5	1.2	1.2	1.8	2.2	2.2	2.2	2.4	2.9
Nuclear power	1.2	2.1	1.9	2.0	1.8	1.8	2	2.1	2.4	3.0
Hydropower	16.4	15.9	14.6	16.7	16.2	14.1	17.2	16.6	18.9	19.5
Other types of power	0.3	0.2	0.3	0.5	1.2	1.6	2.2	2.8	3.3	3.9
Clean energy power	17.9	18.2	16.8	19.2	19.2	17.5	21.4	21.5	24.6	26.4

Note: The data of 2011–15 comes from the China Electricity Council (CEC); the data of other years comes from the IEA and is slightly different from the statistics released by the CEC.

on International Energy Agency (IEA) statistics, in the world's power generation structure of 2013 the proportions of coal-fired, oil-fired, gas-fired, nuclear, hydraulic power, and other types (solar, wind, and biomass) of powers were, respectively, 42.1%, 4.4%, 21.8%, 10.6%, 16.3%, and 5.7%, with clean energy power accounting for 32.6% of the total, where the proportion of coal-fired power was roughly 33 percentage points lower than that of China but the proportions of oil-fired, gas-fired, nuclear, and other types (solar, wind, and biomass) of powers were, respectively, 43, 19.6, 8.5, and 3 percentage points higher than those of China.

As illustrated in Table 2.3, the proportion of clean energy power in the total power generated was near or above 30% in 2014 in most of the major developed countries, except Japan, where the proportion of clean energy power was relatively low due to the shutdown of nuclear power plants. Specifically, the proportion of clean energy power generation was 32.5%, 30.7%, 43.4%, and 94.6% in the United States, South Korea, Germany, and France, but only 24.6% in China because of a low proportion of nuclear and nonhydraulic renewable energy power generation.

The proportion of energy consumed by power generation in total primary energy consumption is 46.2% in France, and above 35% in the United States, Japan, South Korea, and China, as presented in Table 2.4. Primary energy is dominated by coal in China but by oil and gas in developed countries where coal is mainly used for power generation. At present, the proportion of energy consumed by power generation in primary energy is still a bit low in China. The proportion of clean energy used for power generation is more than 85% in Japan, South Korea, and France, and above 75% in the United States and Germany. The proportion of non-fossil energy used for power generation is 86.9% in China, if noncommercial use of biomass energy is not taken into account. The proportion of coal used for power generation in the total consumption of coal is 92.3% in the United States, 83.5% in Germany, 70.2% in South Korea, 59.2% in Japan, and all these percentages are above the Chinese level.

Table 2.3 Power generation structures of some countries in 2014 (%)

Category	United States	Japan	South Korea	Germany	France	China
Coal-fired power	39.8	33.1	4.1	45.2	2.2	70.5
Oil-fired power	0.9	11.2	3	1	0.4	0.1
Gas-fired power	26.8	40.5	24.2	10.1	2.5	2.4
Nuclear power	19.3	0	28.9	16	78.3	2.4
Hydropower	6.1	8	0.7	3.2	10.9	18.9
Nonhydraulic renewable energy power generation	7.1	7.2	1.1	24.2	5.4	3.3
Clean energy power	32.5	15.2	30.7	43.4	94.6	24.6

Note: The data of China comes from the statistics of 2014 released by the China Electricity Council; the data of other countries comes from the IEA, the Energy Balances of OECD Countries 2015, and the Energy Balances of non-OECD Countries 2015.

Table 2.4 The structure of power-generating energy in some countries

Category	United States	Japan	South Korea	Germany	France	China
The proportion of power-generating energy in total primary energy	36.3	40.2	35.8	35.6	46.2	39.1
The proportion of clean energy used for power generation	78.5	85.0	89.8	75.6	90.0	86.9
The proportion of power-generating coal in total coal consumption	92.3	59.2	70.2	83.5	49.6	46.4

Note: *The above data is cited from* IEA Energy Balance Flows. *Power-generating energy includes all primary energy that flows into power stations. The data of proportion of clean energy used for power generation is cited from* IEA Deadline Energy Data. *The data of China is the data of year 2015, which is cited from the* Analysis Report on the Supply and Demand of Power-Generating Energy and the Development of Power Supply in China in 2016.

2.1.3 END-USER ENERGY CONSUMPTION STRUCTURE AND NON-FOSSIL ENERGY UTILIZATION

China's "final energy consumption structure" (End-User Energy Consumption Structure—EUECS) is dominated by coal. With the development of the economy and the energy industry over the years of reform and opening up, the consumption of high-quality energy has been on a steady increase, its proportion in EUECS has been on the rise, but the proportion of coal in EUECS has been on the decline, as shown in Fig. 2.2. In 2015, the proportion of coal, oil, natural gas, electricity, heat, and other energy sources in China's EUECS was 42.4%, 24.0%, 6.2%, 21.3%, and 6.0%, respectively. In particular, clean energy such as electricity and heat accounted for 27.4% of final energy consumption in 2015, but this percentage was only 9% in 1980, showing a significant improvement in EUECS.

In the decade of 1990–2000, China stepped up its pace of EUECS adjustment. The proportion of coal in EUECS fell rapidly from 68.7% in 1990 to 49.6% in 2000, down 19.1 percentage points. In the same period, the proportion of oil in EUECS surged from 17.1% to 27.2%, up 10.1 percentage points; that of natural gas edged up to 2.6%, an increment of 0.1 point; and that of electricity rose from 9.1% to 14.5%, up 5.4 percentage points.

With rapid social and economic development and fast increase in energy consumption, China's EUECS has experienced some fluctuations since 2000. In the period 2000–15, the proportion of coal in EUECS declined from 49.6% to 42.4%; that of electricity rose from 14.5% to 21.3%; that of natural gas increased from

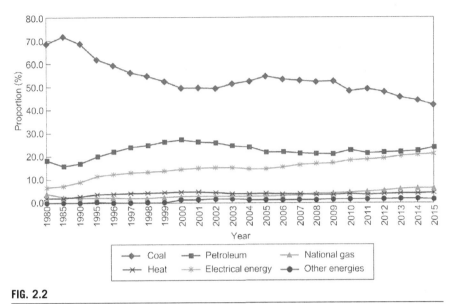

FIG. 2.2

The EUECS of China.

Data from: China Energy Statistics Yearbooks 2014, 2015, and 2016.

2.6% to 6.2%; but that of oil fell from 27.2% to 24.0%, due to high oil prices and limited oil supply capacity.

The world's EUECS is still dominated by oil and natural gas. In 2013, the proportions of coal, oil, natural gas, electricity, heat, and other energy in the world's EUECS were, respectively, 10.4%, 40.4%, 15.3%, 18.3%, and 15.6% (calculated using the data of the IEA, the same below). Compared with the situation in developed countries, the proportion of coal in China's EUECS is quite high, while the proportions of oil, natural gas, heat, and other energy sources are rather low, as shown in Fig. 2.3.

The EUECS of major developed countries is dominated by oil, which is mainly because the proportion of energy consumed by the transportation industry is high in EUECS. In 2013, the proportion of oil in EUECS was close to or above 50% in the United States, Japan, and South Korea, 43.4% in France, and 41.8% in Germany, 1–20 percentage points higher than that in China; the proportion of natural gas in EUECS was above 20% in the United States, France, and Germany, above 10% in South Korea and Japan, but only 5.2% in China; the proportion of coal in EUECS was less than 2% in the United States and France, less than 1% in other major developed countries, but was as high as 33.2% in China (calculated using the data of the IEA); the proportion of electricity in China's EUECS was slightly higher than the world's average, but lower than that of developed countries such as the United States, Japan, South Korea, and France, among which Japan had a proportion as high as 2.2%.

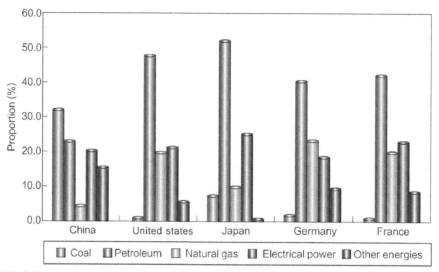

FIG. 2.3

Comparison of some countries' EUECS in 2013.

Data from: IEA, Energy Balances of OECD Countries 2015, Energy Balances of non-OECD Countries 2015.

2.2 MAIN PROBLEMS IN CHINA'S ENERGY DEVELOPMENT

The 18th CPC (Communist Party of China) National Congress set the grand goal of building a moderately prosperous society in all respects by 2020. Realization of the goal calls for the vigorous promotion of ecological civilization and the coordinated development of man and nature. One of the major measures for ecological civilization is to revolutionize energy production and consumption. This puts forward higher requirements for China's energy development. Faced with severe challenges in supply, resources, environment, climate, technology, systems, and so on, the present method of energy development has become unsustainable. The main problems faced by China's energy development are as follows: the capacity for the sustainable supply of conventional fossil fuels is inadequate; the Chinese economy depends more and more on energy imports; the energy transport system is irrational, with increasing pressure on the transportation of power-generating coal; the current power system is difficult to adapt to the rapid growth of renewable energy; there is serious damage to the ecological environment and a mounting pressure on carbon emissions.

2.2.1 RAPIDLY GROWING BUT INEFFICIENT ENERGY CONSUMPTION

2.2.1.1 Continuous and rapid growth of energy consumption

With the rapid development of the economy and the continuous improvement in people's living standards over the years of reform and opening up, China's energy consumption has been increasing markedly. The country's total energy consumption was equivalent to 4.3 billion tons of standard coal in 2015, about eight times more than that in 1978, representing an average annual increase of 5.6%. In the period 2000–15, China's average annual growth rate of energy consumption was 7.4%, far higher than the world's average annual growth rate of 2.2% in the same period. Today, China has become the number one energy consumer in the world.

It is estimated that China's energy demand will keep growing at a fast rate in the years to come. The country is now at the stage of accelerated industrialization and urbanization, which is characterized by the rapid development of heavy industries and a high consumption of energy. On the one hand, large-scale infrastructure construction requires the development of high energy-consuming industries such as iron and steel, cement, and so on. On the other hand, per capita energy consumption in urban living is several times higher than that in rural living, so urbanization will lead to a rapid growth of energy consumption. With the process of accelerated urbanization and industrialization, China will face more severe challenges and a greater pressure in controlling the energy consumption demand. According to domestic and foreign studies, China's primary energy consumption demand will be equivalent to about 50 million tons of standard coal by 2020.

2.2.1.2 The efficiency of energy utilization is relatively low

At present, China's energy consumption per unit of GDP is higher than that of developed countries. In 2015, the figure was 0.635 ton of standard coal per RMB 10,000 of GDP. In comparison with other countries based on PPP (Purchasing Power Parity),[1] China's energy consumption per unit of GDP is about the world's average, similar to that of the United States, or 1.1 times that of Japan. In 2015, China's GDP accounted for only 15.5% of the world's total, but its energy consumption accounted for 18.6% of the world's total energy consumption.[2]

The energy consumption level of main energy-consuming products in China is also higher than the international advanced level. The comparable energy consumption per ton of steel (large- and medium-sized enterprises) is 662 kg of standard coal in China, 52 kg higher than the world's advanced level. Power consumption per ton of electrolytic aluminum is 13,740 kWh, comprehensive energy consumption per ton of ethylene is 879 kg of standard coal, comprehensive energy consumption per ton of synthetic ammonia is 1532 kg of standard coal, and that of paper and cardboard is 1114 kg of standard coal, respectively, 840 kWh, 250, 542, and 510 kg higher than the international advanced level.

There are four main reasons for China's high level of energy consumption. Firstly, energy consumption per unit of GDP in primary industry is higher than that in secondary and tertiary industries. In 2013, the gross product of secondary industry was 43.7% of GDP, a proportion much bigger than that in developed countries, and

[1] Development Planning Department of the National Energy Administration, Development Planning Department of the State Power Grid Corporation, State Grid Energy Research Institute Co., Ltd., Energy Data Handbook 2015.
[2] BP, Statistical Review of World Energy 2015.

the energy consumption of secondary industry accounted for 71.5% of China's total energy consumption. Secondly, there has been no fundamental change in our economic development mode characterized by high investment, high consumption, high emission, high pollution, incoordination, low efficiency, and noncirculation, and the extensive mode of economic growth leads to high-energy consumption and low-energy efficiency. Thirdly, our energy consumption structure is dominated by coal, most of which is directly burnt at the end user's location without proper processing or conversion, thus leading to low efficiency in the energy system. Fourthly, the economy mainly relies on the export of great quantities of cheap goods whose production consumes huge amounts of energy.

2.2.2 ENERGY SUPPLY SECURITY IS INSUFFICIENT

2.2.2.1 Unsustainable supply of conventional fossil energy

Although the total amount of energy resources is abundant in China, the resources of high-quality fossil energy (petroleum and natural gas) are relatively insufficient.[3] Among the proven reserves of mineral energy in China, coal accounts for about 96% while oil and gas account for only 4%. At the end of 2014, the remaining proven recoverable reserves of coal took third place in the world, and those of oil and gas took 14th and 13th places, respectively. Due to the constraints of energy resources, mining conditions, ecological environment, supply sustainability, and other factors, China's sustainable supply capacity of proven conventional fossil energy[4] is equivalent to 3.6 billion tons of standard coal,[5] which includes 4.1 billion tons of coal, 200 million tons of oil, and 300 billion cubic meters of natural gas. Further expansion of production scale will bring about a series of problems such as environmental pollution, production accidents, etc.

The amount of energy resources per capita in China is far below the world average. According to the Energy Data Handbook 2015,[6] China's per capita reserves of coal, oil, and natural gas are, respectively, 65.3%, 5.3%, and 7.7% of the world's average. Yet, the country's per capita energy consumption has been in level with the world's average (in 2013, China's per capita energy consumption reached

---

[3]Natural gas mainly refers to conventional natural gas.

[4]In its 2017 report, Our Shared Future, the World Commission on Environment and Development points out that sustainable development is to meet the needs of the current generation without damaging the needs of the future generation. Under the theoretical framework of sustainable development, the capacity of sustainable energy supply refers to the long-term stable energy supply capacity with comprehensive consideration given to environmental impacts. For example, the sustainable supply capacity of coal needs to consider the constraints of domestic coal resources, geological conditions, ecological environment, water resources, transportation economy, natural disasters, and so on.

[5]Development Planning Department of the National Energy Administration, 2010 Research Report on National Energy Strategy for Scientific Development by 2030.

[6]Development Planning Department of the National Energy Administration, Development Planning Department of the State Power Grid Corporation, State Grid Energy Research Institute Co., Ltd., Energy Data Handbook 2015.

3.1 tons of standard coal/year), and is rapidly growing. China is already a net importer of the three major kinds of fossil energy (oil, natural gas, and coal), and the contradiction between increasing energy demand and limited capacity of energy supply is more and more prominent.

2.2.2.2 Dependence on import energy is on the increase

According to the China Energy Statistics Yearbook 2014, the country's net import of energy increased at an average annual growth rate of 13.6% in the period 2005–13. The amount of primary energy imported by China in 2013 was equivalent to 660 million tons of standard coal, about 16.6% of country's primary energy supply, 150 million tons or 10 percentage points more than in 2005. In the medium and long term, China's energy demand will continue to grow rapidly, and the gap between supply and demand will make the country more and more dependent on energy import.

Specifically, China's coal import volume exceeded its export volume for the first time in history in 2009, becoming a net importer of coal. In 2013 China became one of the world's largest importers of coal, importing 330 million tons of raw coal while exporting only 7.51 million tons, up 710.7% and down 70% from 2008, respectively, with a net import of more than 300 million tons. With respect to crude oil, since China became a net importer of oil in 1993, its dependence on imported oil had increased from 6% to 45% by 2006. The country imported 282 million tons of crude oil in 2013, which represented an external dependence of 57.6%. With regard to natural gas, the country's domestic consumption began to rise rapidly in 2009, especially in the winter when a cold wave hit most parts of the China, thus leading to a sharp increase in gas import. China's dependence on imported gas was 11.6% in 2010 and 29.2% in 2013, an increase of nearly 20 percentage points in just 3 years, and this trend is expected to continue in the years to come. This will bring a series of risks to the country politically, economically, or even militarily.

With the acceleration of economic globalization it is necessary and natural for China to appropriately increase energy imports, but at the same time, we need to establish a new concept of energy security, and reexamine and pay more attention to the issue of energy security. There are two main reasons for this. Firstly, China, as a "catch-up" developing country with very limited control over international oil resources, does not have a strong voice in the affairs of world oil security. As world oil and gas supply is more and more concentrated, the security of energy supply has become the focus of geopolitical struggle in the world. Today, energy is a strategic weapon that affects international political relations, the role of political factors in the competition of oil and gas resources is bigger and bigger, and the political risks in resource-rich regions are on the rise. Secondly, the impact of international energy costs, represented by the price of oil, on energy security or even national economic security is increasing. International petroleum consortiums, by taking advantage of their monopoly over oil resources, have formed an international oil-pricing mechanism based on the oil futures trade, using financial means to intervene in this trade between countries. The violent fluctuations in and repeated surges of the oil price pose a serious threat to the energy security or even economic security of China, a major oil importer and developing country with a huge energy consumption.

2.2.3 IRRATIONAL ENERGY CONFIGURATION

2.2.3.1 Energy transport is dominated by coal, while the proportion of power transmission is relatively low

Coal constitutes the largest part of China's energy sources. Since it is distributed mainly in the northern and western regions of the country, China's energy transport pattern is featured by the "transportation of coal from the north to the south and from the west to the east." For a long time, China's energy transportation has been absolutely dominated by and reliant on coal transportation, while the proportion of power transmission is very low. In 2011, about 1.45 billion tons of coal was directly transported from Shanxi, Shaanxi, and Western Nei Mongol to other parts of the country by railway and highway, yet only 185 TWh of electric power was transmitted from the above three regions, equivalent to a mere 76 million tons of coal, which means that the ratio of coal transport (1.45 billion tons) to power transmission (an equivalent of 76 million tons of coal) is about 19:1, or that the proportion of power transmission in total energy transport is only about 5%, as shown in Fig. 2.4. Transporting a huge volume of coal overburdens the transportation channels, and the utilization ratio of main railway lines for coal transportation has been nearly 100% or even in an overloaded state for many years, posing a serious threat to the security of the energy and power supply.

Coal transportation involves a great many links and the cost fluctuates significantly, which is an important factor for the high price of power-generating coal in coal-receiving areas. Recent years have seen big fluctuations in the coal price. The price of power-generating coal at Qinhuangdao port was RMB 400/ton in 2006, in excess of RMB 800/ton in 2011, but less than RMB 400/ton in 2015. Although the overall price of coal has changed a lot, the cost of transportation

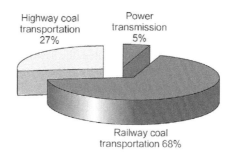

FIG. 2.4

Ways of coal transportation from Shanxi, Shaanxi, and Western Nei Mongol and their corresponding proportions.

has been about 50% of the final price in coal-receiving areas in the country's eastern and central regions. (The price of 5500-kcal power-generating coal in Zhenjiang was RMB 950/ton in 2012 and RMB 520/ton in 2015, yet in Shanxi, the price was RMB 500/ton and RMB 250/ton, respectively.) Due to transportation costs, the coal price in coal-receiving areas has remained at a high level despite a slight decline.

With respect to the layout of coal-fired power generation capacity, an important principle we have followed for many years is that every province and autonomous region should have coal-fired power plants (CFPPs) to meet their own needs. This is a deep-seated reason for the repeated occurrence of coal price hikes in the central and eastern parts of the country. According to this principle, a large number of CFPPs have been built in China's eastern and central regions where the demand for power is huge. At present there is one CFPP every 30 km along the Yangtze River and the distance is even shortened to 1 km in the section between Nanjing and Zhenjiang. Such a density of CFPPs plus huge coal consumption by other industries has led to severe environmental pollution and shortage of land in the eastern and central regions. Worse still, huge amounts of coal have to be transported from the west and the north by highway, railway, or sea, which exerts a great pressure on the country's transportation system and sharply increases the coal price. This development mode has proved unsustainable and the current energy transportation system needs to be rationalized.

2.2.3.2 The existing power system cannot meet the needs of rapid growth of renewable energy

During 12th five-year period, the average annual growth rate of wind power was about 33%. In 2015, China's (except for Taiwan) newly installed capacity of wind power was 32.49 GW. As of by the end of 2015, the grid-connected wind power generating capacity was 128.3 GW. In the future, nine wind power bases, with a 10-GW installed capacity each, will be built in Hami, Jiuquan, West Nei Mongol, East Nei Mongol, Jilin, Heilongjiang, Hebei, Shandong, and coastal Jiangsu. By 2020, the total capacity will reach at least 210 GW. With the rapid large-scale development of renewable energy, the development of wind power is increasingly uncoordinated with that of other power sources and power grids. The system lacks peak-regulating ability, trans-regional power transmission ability, and other capabilities, such as, restricting the development scale and utilization efficiency of wind power and other renewable energies. Therefore, we must strengthen the unified planning of systems and speed up our research to solve related technical problems as quickly as we can.

From the perspective of power structure, in the northeast, north, and northwest of China (the "three northern areas") where wind resources are abundant, the power structure was still dominated by thermal power in 2015. The proportion of coal-fired power was 72% in the northeast and 76% in the north, showing a severe shortage of flexible and efficient peak-regulating power supply, as illustrated in Fig. 2.5. At present, power sources with a good peak-regulating ability such as pumped-storage power stations and gas-fired power stations only account for less than 3% of the total installed capacity, far below the level of 30%–50% of developed European and North

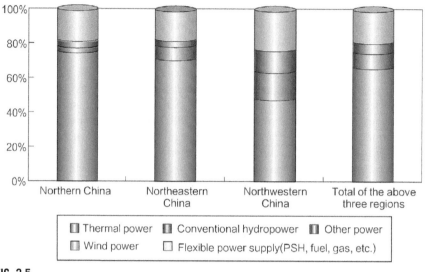

FIG. 2.5

Power structure in the north, northeast, and northwest of China (the "three northern areas").

American countries. This situation makes it difficult for the present power system to accept and efficiently utilize wind power.

From the perspective of power grid scale, China's trans-provincial and trans-regional power transmission capacity is 110 GW at present, about 19.5% of the maximum load of the whole society, far below the level of developed European and North American countries where wind power constitutes a significant part of their power sources. For example, the electricity exchange capacity between Denmark and its neighboring countries accounts for as much as 85% of the total load of the whole society. At present, China's installed capacity of wind power accounts for only 8.5% of the national total of installed power generating capacity. The central and eastern regions are rich peak-regulating resources, yet their market potential for accommodating wind power is not fully utilized. From the perspective of development, grid construction lags behind and in particular, trans-regional transmission capacity is insufficient, which will directly affect the large-scale development and efficient utilization of wind power and other renewable energies. Fig. 2.6 shows the situation of interconnection of power grids in China (by the end of 2014).

2.2.4 THE SITUATION OF ENERGY AND ENVIRONMENT IS GRIM

2.2.4.1 Ecological environment is seriously damaged

Intensified mining of coal resources has caused serious environmental problems such as land subsidence and water pollution. Firstly, overexploitation of coal has led to land subsidence and destruction of the ecological environment. According to

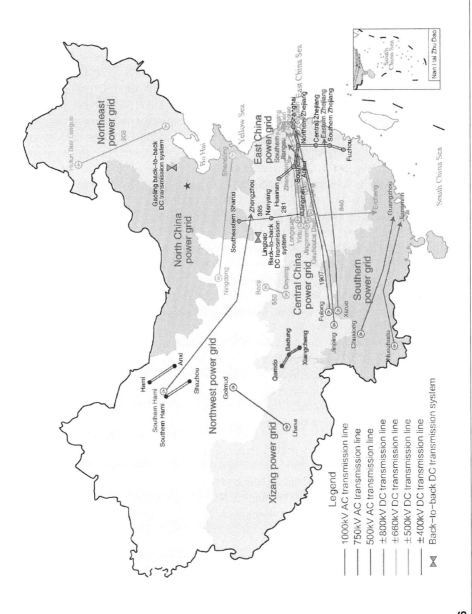

FIG. 2.6

Nationwide interconnection of power grids in China (by the end of 2014).

statistics, the total area of surface subsidence caused by coal mining has reached 800,000 ha in recent years, and it is increasing at a rate of about 40,000 ha per year. The reclamation rate of land damaged by coal mining is only 25%, far below the average level of 65% in developed countries. Ground subsidence will affect or even destroy buildings, roads, land, and rivers, which is an important factor constraining coal mining in the central and eastern part of China. Secondly, coal mining has resulted in water pollution. Coal production will inevitably affect the geological conditions of groundwater and cause the decline of the groundwater level. In China, most of the regions rich in coal resources have a severe shortage of water. Of all the major mining areas, 71% are located in water-deficient regions and 40% in severely water-deficient regions. Water resources are seriously damaged in the main coal-producing areas such as Shanxi, Shaanxi, and Western Nei Mongol. Every year huge amounts of wastewater is pumped out from coal mines and directly discharged into the environment, polluting the surface water and groundwater. The destruction of water resources and the pollution of the ecological environment have become the main constraint factors for coal development in Central and Western China. Thirdly, solid wastes from coal mines occupy land and pollute groundwater, soil environment, and atmospheric environment. At present, coal gangue is one of the largest kinds of solid wastes in China, where mining each ton of coal will generate about 0.13 ton of coal gangue, whose comprehensive utilization ratio is only about 61%,[7] far below the level of 90% in the developed countries. Huge piles of coal gangue not only occupy land, but also cause serious air pollution and water pollution due to spontaneous combustion and leaching of the gangue. Gas released from coal mining is also a main source of greenhouse gas emissions.

The coal-based energy structure, the direct combustion of a large amount of coal in terminal equipment, and the exhaust emissions from an increasing number of motor vehicles have caused serious composite air pollution in China. Coal combustion is responsible for 90% of sulfur dioxide emissions, 67% of nitrogen oxide (NOx) emissions, 70% soot emissions of the country, and 25%–45% PM2.5 concentration in provincial capital cities in the eastern and central regions.[8]

(1)   China ranks first in the world in terms of sulfur dioxide and nitrogen oxide emissions, and its acid rain is serious in some areas. In 2013, acid rain affected about 1 million km2 of land in which 60,000 km^2 were severely affected. Acid rain mainly occurs in the Yangtze River Delta, the Pearl River Delta, and other regions where energy consumption is highly concentrated. In recent years, the proportion of the power industry's sulfur dioxide emissions in the

---

[7]Work Report at the Fourth Session of the Fifth Council of China Coal Processing & Utilization Association, http:/www.ccpua.org/article/detail,asp?articleld=905.
[8]Data of sulfur dioxide, nitrogen oxides, and soot: State Power Grid Corporation, Company Value, 2012; data of PM2.5: Tsinghua University, State Grid Energy Research Institute Co., Ltd., The Influence of Future Air Pollution Control on the Layout of Electric Power Industry, 2012.

national total has fallen from 51% in 2005 to 35% in 2014 thanks to the vigorous promotion of all kinds of emission reduction measures.

(2) PM2.5 pollution is quite serious in Beijing, Tianjin, Hebei, the Yangtze River Delta, and the Pearl River Delta regions. According to pilot monitoring results in 2013, smoggy days in the most seriously affect areas of these regions accounted for more than half of the year. Although the power industry's coal consumption is as much as 50% of the total social consumption, its contribution to PM2.5 pollution has been reduced through de-dusting, desulfurization, denitrification, and other comprehensive measures. Considering the impacts of primary and secondary PM2.5,[9] the power industry's contribution to PM2.5 concentration in the provincial capital cities in the east and central regions was 7%–21% in 2013.[10]

(3) The problem of mercury emissions caused by coal combustion has also attracted wide attention. Up to 2010, China's annual mercury emission was 500–600 tons, about one-fourth of the global total. Specifically, mercury emission from coal combustion is responsible for 40% of artificial-source atmospheric mercury emissions,[11] posing a serious threat to the living environment of human beings.

2.2.4.2 The pressure of carbon emission reduction is increasing

With the rapid development of the economy and the continuous growth of energy demand, carbon dioxide emissions from energy activities are also increasing in China. Since 2003 in particular, China's fuel demand has been on the rise due to accelerated economic growth. At the same time, carbon dioxide emissions from energy activities increased rapidly at an annual growth rate of 10%. In 2014, the country's carbon dioxide emissions from fossil fuel combustion was about 9.76 billion tons,[12] compared with the United States' 5.99 billion tons.[10] In 2014, China's per capita carbon dioxide emission was 7.1 tons, higher than the global average of 4.9 tons.

China is faced with a much greater pressure than developed countries in dealing with climate change. Our current economic and social development stage determines that our energy demand will see a great increase in the future, and our coal-based energy structure determines that China's coal consumption will also increase significantly. When providing an equal amount of energy, coal produces 29% and 69% more carbon dioxide than oil and natural gas, respectively, which means that we face

---

[9]The PM2.5 directly emitted into the atmosphere from CFPPs is referred to as primary PM2.5, while the sulfates and nitrates transformed from sulfur dioxide and nitrogen oxide emitted from CFPPs are called secondary PM2.5.

[10]Tsinghua University, State Grid Energy Research Institute Co., Ltd., The Influence of Future Air Pollution Control on the Layout of Electric Power Industry, 2012.

[11]Progress of Research on Mercury Emission from Coal Combustion Flue Gas and Control Thereof, Journal of Safety and Environment, 2012, 02.

[12]BP, Statistical Review of World Energy 2015.

a much greater pressure in reducing greenhouse gas emissions than developed countries where the industrialization process has been finished and where the energy structure is based on oil and gas. Considering that by 2020 developed countries are expected to reduce carbon dioxide emissions by 20% from the level of 2005, if China does not take effective measures immediately, its increment in carbon dioxide emission from the present year up to 2020 will exceed the total emission reduction in developed countries in the same period. So, China will face more and more international pressure with regard to the issue of global climate change.

2.3 IDEAS ABOUT FUTURE ENERGY DEVELOPMENT IN CHINA

The Outline of the 13th Five-Year Plan for National Economic and Social Development of the People's Republic of China ("the Outline of 13th Five-Year Plan" for short) proposed the strategic idea of "pushing forward energy revolution, transforming the way of energy production and utilization, optimizing the energy supply structure, improving energy utilization efficiency, and building a modern, clean, safe, efficient, and low-carbon energy system to ensure national energy security." According to this idea we should, on the one hand, strengthen energy saving as a priority and control the total amount of consumption, and on the other hand, promote energy diversification, strive for clean development, and optimize the energy structure and development layout.

2.3.1 CONTROL OF TOTAL ENERGY CONSUMPTION (CTEC)

2.3.1.1 The proposal of control of total energy consumption (CTEC)

As an important measure to improve the incentive and restraint mechanism of energy saving and emission reduction, "reasonable control of total energy consumption" has been included in the Outline of 13th Five-Year Plan. The main background and ideas of the CTEC policy described are as follows.

First, although China is now at the stage of rapid economic growth, the mode of economic development is extensive, leaving a big space for energy saving. By CTEC, we can reduce unreasonable energy demand, eliminate backward production capacity, and lower the cost of energy demand. Second, by CTEC we will be able to control carbon emissions from the source. The CTEC is mainly targeted at fossil energy whose consumption is the main source of carbon emissions. The carbon emissions generated from fossil energies account for more than 90% of the world's total. Third, while continuing to take energy intensity as a constraint, we should carry out CTEC and tighten assessment of energy saving and emission reduction. We have been taking energy consumption per unit of GDP as a constraint index, but it does not sufficiently restrain energy consumption although it is conducive to economic growth.

2.3.1.2 Goals and methods of CTEC

As prescribed in the Outline of 13th Five-Year Plan, the total national annual consumption of energy should be controlled within an equivalent of 5 billion tons of standard coal by 2020. With "determining base amount and decomposing increment" as the basic idea and method, China divides its provincial-level regions (except Xizang) into five groups according to the level of economic and social development, the characteristics of energy consumption, the endowment of energy resources, regional policies, and some other factors. Different groups correspond to different growth rates of energy consumption; the national total consumption of energy and electricity is distributed to provincial-level governments. At the same time, energy saving should be promoted in industries, buildings, transportation, public institutions, and other fields by upgrading boilers (kilns), lighting devices, and motor systems, and using residual heat for residential heating. Key energy-consuming enterprises should take drastic actions, carry out voluntary energy-saving activities, promote the construction of energy management systems, measurement systems, and energy consumption online monitoring systems, and conduct energy evaluations and performance evaluations. Efforts should be made to improve the energy efficiency of buildings, formulate a whole-industry-chain plan for the development of green buildings, implement energy saving and low-carbon power dispatching, and promote a cascaded comprehensive utilization of energy.

2.3.1.3 Upgrading the level of electrification is an important handle for CTEC

The CTEC will enhance the level of electrification and the latter will, in turn, promote the realization of CTEC goals. Firstly, CTEC will drive the adjustment and optimization of industrial structure, by which the proportion of energy consumed by residents and tertiary industry will be increased in the nation's total energy consumption. As the electrification level of tertiary industry and peoples' lives is higher than that of secondary industry, CTEC will bring up the overall electrification level of China. Secondly, CTEC will speed up the development of renewable energy and also help to improve the level of electrification. Thirdly, the rise of the electrification level helps improve energy efficiency and promote the realization of CTEC goals. Statistics show that in the recent 30 years China's electrification level has a negative correlation with energy consumption intensity: every 1% increase in the proportion of electrical energy in EUECS will lead to a decrease of 3%–4% in the energy consumption per unit of GDP.

2.3.2 ENERGY STRUCTURE ADJUSTMENT AND LAYOUT OPTIMIZATION

China should accelerate the development of natural gas, vigorously tap hydropower, steadily develop nuclear power, efficiently develop wind power, expand the utilization of solar energy, and increase the proportion of high-quality fossil energy and clean energy so as to realize the adjustment of energy structure. In the future, China's capacity of primary energy supply will mainly be from the central and western regions. For this reason, we need to intensify energy development in the western region and enhance the capacity of trans-regional energy transmission; we need to stabilize the intensity and rhythm of energy development in the central region and maintain its strategic role in linking the eastern and western regions; we also need to optimize the eastern region's energy development so as to form an energy development layout where the advantages of the eastern, central, and western regions are linked with each other in an orderly and complementary manner.

2.3.2.1 The development of coal and CFPPs

According to the Strategic Action Plan for Energy Development (2014–2020), China's total consumption of coal will reach the peak, 4.2 billion tons per year, accounting for 62% of the total consumption of primary energy.

In accordance with the principle of "controlling the eastern region, stabilizing the central region, and developing the western region," we should strengthen the intensive development of coal resources in a safe and green manner. We should speed up the construction of six coal bases in Northern Shaanxi, Huanglong, Shandong, Eastern Nei Mongol (northeast), Ningdong, and Xinjiang; optimize the development of coal resources in Northern, Central, and Eastern Shanxi, as well as in Yunnan and Guizhou; stabilize the scale and intensity of development of coal mines in Central Hebei, Western Shandong, Henan, Huainan, and Huaibei. In the future, new capacity for coal production will be mainly distributed in Nei Mongol, Shanxi, Shaanxi, Ningxia, and Xinjiang, meaning that the center of coal production will gradually move westward and northward.

We will strictly control the total number of CFPPs, especially those in the central and eastern regions. The absolute majority of new CFPPs will be built in the western and northern regions. The Action Plan for the Prevention and Control of Air Pollution requires us to strictly control the construction of new CFPPs in Beijing, Tianjin, Hebei, the Yangtze River Delta, and the Pearl River Delta regions, and forbids the approval of new coal-fired power generation projects except for heat and power cogeneration. During the 13th five-year period, China will build nine 10-million-kW class large-scale coal-fired power generation bases in Xilingol, Ordos, Northern Shanxi, and other locations, with new CFPPs mainly distributed in northern and western regions. It is expected that by 2020 the national total installed capacity of CFPPs will reach 1.12 billion kW, about 70% of which will be new capacity installed in the period 2015–20 in Shanxi, Shaanxi, Nei Mongol, Ningxia, and Xinjiang. The proportion of installed capacity in eastern and central regions will decline from 46% in 2014 to 36% in 2020, down by 10 percentage points.

2.3.2.2 Natural gas and gas-fired power generation

With the acceleration of natural gas exploration, development, and importation, its proportion in China's PECS will increase steadily. As an important part of natural gas consumption, appropriate development of gas-fired power generation and distributed energy systems with natural gas as fuel is beneficial for the improvement of the power supply structure and the clean and efficient utilization of energy.

We need to appropriately develop large-scale gas-fired power stations and promote gas-based distributed energy supply systems. According to National Gas Utilization Policies, natural gas should be firstly used in urban residents' living and public service facilities before being used for other purposes. In the 13th five-year period, gas-fired power generation should play two roles. The first is to adjust the structure. In key areas for air pollution prevention and control such as Beijing, Tianjin, Hebei, Shandong, the Yangtze River Delta, and the Pearl River Delta, gas-steam combined cycle cogeneration should be appropriately developed according to thermal load demand. The second is to promote the development of new energies. In areas where the development scale of new energies is relatively large, we should appropriately develop single-cycle units so as to improve the system's peak regulating ability. Considering the security level, external dependency, and price trend of natural gas supply, it is suggested that China's gas-fired power generation capacity reach 85 GW, an increase of 30 GW from 2014.

2.3.2.3 Hydropower development

Since China is rich in hydropower resources and our hydropower technology is relatively mature, giving priority to the development and utilization of green and renewable hydropower resources is an important support for the sustainable development of energy in China. With respect to hydropower, the Outline of 13th Five-Year Plan requires us to "coordinate hydropower development and ecological protection, give priority to ecological protection, and scientifically tap hydropower resources in Southwest China by focusing on the construction of major hydropower stations in important watersheds."

The development of hydropower bases is gradually shifting westward and advancing step by step. Before 2020, the key areas for hydropower development will be mainly concentrated in Sichuan and Yunnan. The construction of eight hydropower bases in the upper reaches of the Yangtze River, Wujiang River, Nanpan River, Hongshui River, the upper reaches of the Yellow River and its northern trunk stream, Western Hunan, Fujian, Zhenjiang, Jiangxi, and Northeast China will be accelerated and completed as soon as possible. Other hydropower bases will be laid out on the Jinsha River, Yalong River, Dadu River, Lancang River, and Nujiang River. By 2020, the development of all high-quality hydropower resources across the country (excluding Xizang) will have been completed. Large-scale development of hydropower will be started in Xizang after 2020.

Hydropower development and transmission should ensure the balance between supply and demand and follow the principle of efficient utilization. Considering the factors such as the direction of energy flow, the substitution rate of hydropower, the reduction of surplus water, and the routes of power transmission, southwest hydropower will mainly be sent to East China, Central China, and South China power grids. From the perspective of long-term sustainable use of transmission routes, the development and transmission of southwest hydropower should be considered in a much broader picture that includes the northwest so as to form a pattern where the CFPPs in Xinjiang and the hydropower stations in Xizang are in coordination and complement each other. We should ensure the overall balance between supply and demand, give full play to the capacity benefits and coal-saving benefits of hydropower, replace as much as possible the capacity of CFPPs in power-receiving areas with hydropower, make efficient use of hydropower in these areas while meeting the needs of power sending areas, so that hydropower stations do not have surplus water in the wet season, and power-sending and receiving areas do not suffer power shortages in the dry season.

2.3.2.4 Nuclear power development

China will efficiently develop nuclear power on the basis of ensuring safety. In the long run, as a big energy-consuming and carbon-emitting country, China imperatively needs to develop nuclear power due to the rigid demand of economic growth and environmental protection as well as the challenge of climate change. By the end of 2014, the country's installed capacity of nuclear power had reached 20.08 GW, with another 35.88 GW under construction. Considering the long construction cycle of nuclear power plants (5–8 years) and the current progress of nuclear power construction, the installed capacity of nuclear power is expected to reach 52 GW by 2020, slightly lower than the target of 58 GW put forward in the Strategic Action Plan for Energy Development (2014–2020). In the next few years, construction of more nuclear power projects must be commenced to ensure that by 2020 the capacity under construction will reach 30 GW at least. As for the layout of nuclear power, China's new installed capacity of nuclear power constructed in the 13th five-year period will mainly be distributed in the eastern coastal region where electrical load increases rapidly and energy sources are scarce.

2.3.2.5 Wind power development

For the sake of sound development of wind power, we must reverse our previous approach of blindly pursuing the increase in the installed capacity of wind power. Based on a careful consideration of the ability of accommodation, the focus of wind power development must shift from speed to quality, from merely pursuing installed capacity to efficiently utilizing such capacity, and from large-scale centralized development to a pattern where large-scale centralized development is supplemented by decentralized development.

Centralized development should take place mainly in areas rich in wind energy, while decentralized development should be carried out mainly in other areas. China's potentially exploitable onshore and offshore wind energy at an altitude of 50 m is at least 2500 GW, which is mainly distributed in the "three northern areas" and the eastern and southeastern coastal areas where the conditions for centralized large-scale development of wind power are available. Large-scale wind power bases will be built in these areas in an orderly manner and the power will be transmitted to other provinces or autonomously regions of the country to expand the range of accommodation. China's several 10-GW class wind power bases, especially those in the "three northern areas," need trans-regional transmission for highly efficient accommodation. In addition, inland areas with abundant wind resources such as hilly and river valley areas are encouraged to build decentralized small- and medium-sized wind power projects based on their local conditions by giving full play to their advantages of being close to power load and having good access to power grids.

2.3.2.6 Solar power development

China's solar photovoltaic (PV) power generation should be developed according to the idea of combining "large-scale centralized development, medium-high-voltage transmission" with "decentralized development, low-voltage local accommodation." In the remote areas of Xizang, Qinghai, Nei Mongol, Xinjiang, Ningxia, Gansu, Yunnan, and other provinces (autonomous regions), the use of household PV power systems and small PV power stations will be promoted to make electricity available for people without a power supply. In the central and eastern regions, we should promote the use of distributed grid-connected PV power generation systems that are combined with buildings, encourage the installation of PV power generation systems in urban public facilities, commercial buildings, and industrial parks where appropriate, and support the installation of grid-connected PV power generation systems on the roof on industrial buildings. We will build large-scale PV power stations on the Gobi Desert in the western/northern provinces such as Qinghai, Gansu, Xinjiang, Xizang, and Nei Mongol, which are rich in solar energy resources and have good conditions for the construction of PV projects.

Following the idea of "demonstration-based promotion," China will build demonstrative PV power projects on wild land, Gobi Desert, and wasteland in Nei Mongol, Gansu, Qinghai, Xinjiang, and Xizang.

2.3.3 CONSTRUCTION OF A COMPREHENSIVE ENERGY TRANSPORTATION SYSTEM

In a traditional sense, a comprehensive energy transportation system is one comprising various transportation modes such as railway, highway, waterway, pipeline, and aviation, as well as related stations, ports, and terminals. The successful application of ultrahigh-voltage (UHV) AC, and DC transmission technology has promoted power grids to be integrated into the comprehensive energy transportation system. Accelerating the development of trans-regional power transmission will play an increasingly important role in promoting energy structure adjustment, rationalizing the allocation of environmental resources, and improving the level of supply security. In recent years the State Grid Energy Research Institute Co. Ltd. has carried out continuous rolling research on the comprehensive comparison between coal transportation and power transmission, achieving some important research results.

Power transmission brings much more social and economic benefits than coal transportation. First, the current spot price of electricity brought to the receiving end through long-distance UHV AC and DC transmission is, in most cases, RMB

0.01–0.07/kWh less than the price of electricity generated by the local CFPPs at the receiving end, so the former has economic advantages and will help to curb the excessive rise of electricity prices. Second, the energy utilization efficiency of "coal transportation plus power transmission" is 0.2%–0.3% higher than that of pure coal transportation. Third, power transmission alleviates the environmental pressure in the eastern and central regions where the power load is concentrated, promotes the optimized utilization of national environmental resources, and reduces national environmental loss. Fourth, the power transmission corridor is "an airborne express-way for energy transportation," the land under which can be utilized, thus saving a large amount of land. Power transmission uses only 1/4–1/2 of land that coal trans-portation needs to use. Fifth, power transmission is more conducive to the economic development in Western China than coal transportation. In Shanxi, for example, power transmission contributes six times more to local GDP and creates twice the number of jobs than coal transportation. Sixth, promoting both coal transportation and power transmission and improving the capacity of power transmission will effec-tively reduce the pressure on railway systems, enable the two forms of energy trans-fer to complement each other in risk control and disaster relief, and help to ensure the security of energy supply in the central and eastern regions.

"Promoting both coal transportation and power transmission while speeding up the development of power transmission" is a key measure to improve the security of energy and power supply in the central and eastern regions of China. In the foresee-able future coal will remain the dominant energy source in the country. For this rea-son, we will build large-scale coal-fired power generation bases near coal mines in the northern and western regions where coal resources are abundant, step up the con-struction of UHV transmission channels, promote both coal transportation and power transmission, improve the comprehensive energy transportation system, and enhance the security level of energy and power supplies in the central and eastern regions.

Speeding up the development of trans-regional power transmission will promote the intensive development and efficient utilization of clean energy, significantly expanding the development scale of renewable energy bases and the accommodation range. In the future, the development of wind power in China will be concentrated in the "three northern areas," yet the load level of power grid in these areas is low, the system size is small, and the proportion of cogeneration units with poor regulating ability is large, which makes the local wind power accommodation capacity insuf-ficient. To achieve large-scale development of wind power, we must transmit it to other regions through trans-regional power grids. As the country's main power load centers with large-scale systems and relatively strong power-regulation ability, North China Power Grid (covering Beijing, Tianjin, Hebei, and Shandong), East China Power Grid (covering Shanghai, Jiangsu, Zhejiang, Fujian), and Central China Power Grid (covering Henan, Hubei, Hunan, Jiangxi) are important markets for accommodating the wind power generated from the "three northern areas." There-fore, the construction of robust trans-regional power grids will help achieve the com-plementation of power supply characteristics of different regions and improve the systems' accommodation ability for renewable energies.

To sum up, in the future we need to speed up the construction of a robust power grid configuration platform that connects large coal-fired power, hydropower, nuclear power, and renewable energy bases, power transmission channels of major load centers, and the areas around these centers. We need to promote the application of high-capacity long-distance UHV transmission technologies, improve the capacity of large-scale trans-regional power configuration and optimization, and significantly increase the proportion of power transmission in China's trans-regional energy transportation.

Methods for and models of research on non-fossil energy development

This chapter, under the framework of the research method system for energy and power, introduces the thoughts on, methods for, and models of non-fossil energy development. It focuses on, and establishes models, for analysis on the coordinated energy and power development, in line with non-fossil energy development goals, and methods for the comprehensive evaluation of energy and power development scenarios in the future, from the aspects of the economy, society, and the environment.

3.1 OVERALL ANALYSIS THOUGHT AND METHOD

Since the 21st century, a new wave of large-scale clean energy development characterized by the faster development of new non-fossil energy sources, including wind power, has started across the world. In such a context, issues such as the feasibility and development path toward higher consumption of non-fossil energy are extensively studied in the world, focusing on raising the proportion of non-fossil energy to total energy use, optimization and expansion of installed power capacity and a grid interconnection plan, and creating a positive policy mechanism and environment for economic rights and interest relationships between major stakeholders within the energy and power industrial system. The Chinese government has put forward thoughts on the framework of non-fossil energy development. However, it has to address such major issues as the development path toward achieving the target of making non-fossil energy account for 20% of the total energy consumption of the country by 2030, controlling the costs of energy and power supply, driving the development of strategic emerging industries, as well as coping with the challenges of the ecological environment and climate change.

Sequentially, this book draws on foreign experience, analyzes relevant problems for China, establishes a model for coordinated energy and power development that is in line with China's national conditions, establishes, optimizes, and adjusts the scenarios for non-fossil energy development, charts a course, and puts forward policies and measures for non-fossil energy development. The overall framework of the research model is shown in Fig. 3.1. First, it looks back on the history of energy development in the world, systematically analyzes the implementation and

Non-Fossil Energy Development in China. https://doi.org/10.1016/B978-0-12-813106-0.00003-9

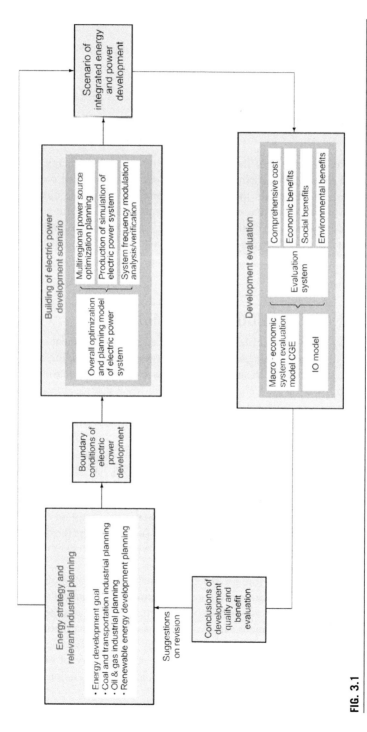

FIG. 3.1

The overall framework of the research model.

adjustment of energy strategies of major developed countries and regions including the United States, the European Union, and Japan, and provides references for China to raise its proportion of non-fossil energy through in-depth analyses on the experience and tendency of clean energy development of foreign countries. Second, it conducts a thorough analysis on the problems of China in terms of total consumption growth, efficiency, supply, allocation, and environment of energy, as well as the principles and thoughts on China's energy development for the future, from the aspects of controlling total energy consumption, making structural adjustments, optimizing energy allocation, and building a system for integrated energy transportation, on the basis of China's national strategy of energy development. The research on the domestic and overseas energy situation will serve as the basis for building China's non-fossil energy development scenario in the future.

According to the basic understanding that non-fossil energy sources are mainly converted into electricity, and on the basis of China's basic national conditions, models of coordinated energy and power development for achieving non-fossil energy development goals are established through research. The models, with the software for overall optimization of the power system at the core, incorporate clean energy utilization into overall power development planning, integrate system tools for power source planning, production simulation, and frequency regulation analysis, and establish a complete toolkit for planning and operation analyses on a clean energy and power system. In addition, a method has been created for comprehensive evaluation of future energy and power development scenarios in terms of economy, society, and environment, by using energy-environment-economic evaluation models, including the computable general equilibrium (CGE) model and the energy development quality evaluation indicator system.

On the basis of establishing the above core method and models, the future policy environment for China's non-fossil energy development, as well as relevant national and local plans, possible development scenarios are built by comprehensively considering resource conditions, supply and demand situations, development progress, and construction cycles of non-fossil energy of China in the future, for the purpose of gaining a deep insight into the laws of coordination between non-fossil energy and fossil energy, and between the power system and the economic environment. The development scenarios must meet the target of China's non-fossil energy accounting for 20% of total primary energy consumption by 2030. Under different scenarios of non-fossil energy development, contrastive analyses are conducted through further evaluation of energy development on the development quality, development benefits, and techno-economic efficiency. On the basis of scenario building and selection, this chapter demonstrates the reasonable path toward and necessary conditions for the realization of the target of 20% non-fossil energy use, covering arrangements, construction cycles, equipment manufacturing capabilities, and talent demands of key projects, for the development of non-fossil energy sources including hydropower and nuclear power, and puts forward policies and measures for achieving the above-mentioned necessary conditions. The specific scenario analyses and evaluation consists of the two aspects below:

(1) Analysis on energy and power development scenarios. Establishes power development scenarios in line with China's energy strategy, industrial planning, and regional planning by making overall plans and coordination for a balance between supply and demand of energy sources such as coal, oil and gas, and electric power and inputting the scenarios as boundary conditions into the model for overall optimization and planning of the power system, consisting of power planning, production simulation, and frequency regulation analysis, for the purposes of minimizing the total cost of power supply, meeting the predicted power demand, and optimizing the power flows from the sending-end energy bases to the receiving-end provinces (autonomous regions) and the power source structure. Synthesizes boundary conditions of energy development create energy and power development scenarios. Analyzes the power structure and distribution, mode of system operation, as well as non-fossil energy power generation and consumption under different development scenarios, based on those of non-fossil energy and power development.

(2) Quality and benefit evaluation of energy development. By using the macroeconomic system model, namely the multiregional dynamic CGE model and the input-output (IO) model, it analyzes the investments, fixed costs of operation and maintenance, variable cost of fuel, external environment costs, as well as economic, energy-saving, resource, environmental, safety, and social benefits brought by input and output of China's future economic and social development under various energy and power development scenarios. It evaluates the development quality of various scenarios, provides the basis for the choice of the path to non-fossil energy development goals, and provides suggestions on the development of energy strategy and relevant industrial plans.

3.2 METHODS FOR OVERALL OPTIMIZATION AND PLANNING OF THE NON-FOSSIL ENERGY AND POWER SYSTEM

Thoughts are on establishing the model for overall optimization of a non-fossil energy and power system, in accordance with the development potential and conditions of coal power, hydropower, and wind power energy bases, with the goal of minimizing the total social costs within the planning period, with the constraints of energy supply capacity, electric quantity balance, peak shaving balance, system operation, and environmental space, design optimization problems and to seek a solution for power generating energy, covering the development scale and time sequence of non-fossil energy power generation, cross-provincial, cross-regional power transmission scales, target markets, as well as installed capacity and distribution of various power sources.

On the basis of analyses on the current situation of China's utilization of power generating energy and the prospect of energy development and utilization, the development scale and distribution of China's hydropower and nuclear power have been basically defined. In the future, most wind power development activities of China will be carried out at the end of the system, where the wind power consumption capacity is limited and the development scale of wind power is mainly constrained by the wind power consumption capacity of the system. In the research, the future development scale and distribution of China's hydropower, nuclear power, wind power, and solar energy have been given, and detailed studies are made on the direction and scale of power flow, as well as the development scale and distribution of coal-fired power stations, pumped storage power stations, and gas-fired power stations.

3.2.1 OVERALL FRAMEWORK OF THE RESEARCH METHOD

Generally, the research method consists of three parts: ① optimize and determine the scale and distribution of power generation based on non-fossil energy and other energy sources by the target year through the generator of electric system planning (GESP-III) model for multiregional power source expansion and optimization. The development plan determined by power source optimization planning is used as the input conditions for calculations in the renewable energy consumption plan. ② In accordance with the defined plans, use the whole-year hourly power system production simulation, including the wind power model to optimize and determine the operating mode of systems for generating power based on energy sources including renewable energy, plans on consumption of renewable energy sources including wind power, as well as wind/solar curtailments. Meanwhile, through production simulation, the output results such as the operating position and fuel consumption level of various units can be obtained. ③ Further check the system production simulation results through frequency regulation analyses, making clear the impacts of the plans on development and consumption of random and intermittent power sources, including wind power, on system frequency stability, and evaluate the applicability of system frequency. Meanwhile, minute-level frequency regulation analyses are conducted on the periods during which the system frequency exceeds the limits, and the wind power consumption plan and wind power curtailment are revised in accordance with the wind power farm curtailment plan and time during the period of difficult frequency regulation.

The calculation method and process are shown in Fig. 3.2.

Under different development scenarios, this method, by adjusting the cross-provincial, cross-regional grid expansion scale, power source optimization and expansion, overall arrangements and scheduling, as well as iterative computations of random production simulation. It may also help to automatically optimize power source planning, methods of system operation, and plans on consumption of renewable energy sources, including wind power, under corresponding boundary conditions, and achieve the goal of non-fossil energy power generation in a relatively economical manner. The main characteristics of the method include:

(1) Multiregional planning. China has many power sending and receiving ends. State Grid Corporation of China (SGCC)-operated power sending-end regions

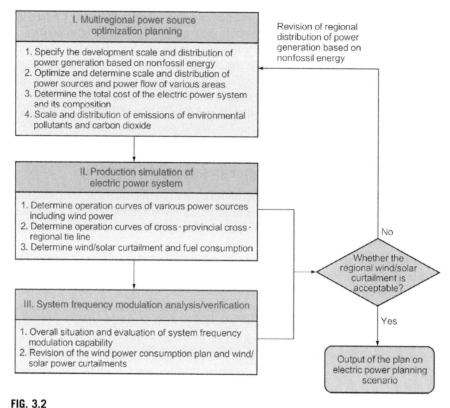

FIG. 3.2

The calculation method and process.

include Shanxi, Shaanxi, Nei Mongol, Ningxia, Xinjiang, Heilongjiang, and southwest China (Sichuan, Chongqing, and Xizang), while power receiving-end regions include Beijing, Tianjin, Hebei, Shandong, the four provinces in Central and East China, and the five provinces (municipalities) in East China. After determining the power transmission capability of the power sending-end regions and the market potential of the power receiving-end regions, the scale and direction of energy and power transmission from the power sending-end regions, as well as the scale and direction of power transmission of power receiving regions, are defined through solving the problem of multiregional power optimization and planning.

(2) Cross-regional grid expansion planning. China is a country with a vast territory, and its primary energy and industrial distribution are uneven. This method considers the geographical distribution of power sources, carries out power source expansion planning, synchronously completes the cross-regional grid expansion plan, and carries out overall optimization of the power system according to the different geographical distribution of power sources and loads.

(3) Non-fossil energy power generation planning. Unlike conventional power units characterized by controllable output, the power generation based on non-fossil energy sources, including wind energy, is random and intermittent, putting forward higher requirements on the method for power system planning. This method, on the basis of different simulation requirements on planning, production simulation and frequency regulation verification, models power generation based on new energy sources, including wind power, on different time scales, and includes them in the power planning model system, thereby incorporating the non-fossil energy power generation planning in the overall power system planning system.

3.2.2 IMPROVING THE SOLUTION-SEEKING PROCESS

(1) Improving plans for the electric power system, including plans on power generation based on non-fossil energy and other energy sources by the target year, and cross-regional power grid plans. The planning process for multiregional power source optimization is shown in Fig. 3.3. The first step is to input the electric power system plan to work out boundary conditions and parameters (see Appendix II), including electricity demand and load characteristics, scale and direction of existing and specified cross-provincial and cross-regional power transmission, specified power sources, techno-economic indicators of various power sources, etc. The next step is to devise the initial power source and cross-regional power grid plans, specifying power source structure, distribution, cross-regional power flow, etc., to achieve the lowest cost of power supply, without regard to large-scale wind and solar power generation, using the GESP-IV (see Appendix III) model for multiregional power source expansion and optimization. On the basis of that, the initial power flow plan should be checked and adjusted, considering the power distribution demand, when there is large-scale wind and solar power generation. The adjusted plan should be used as a boundary condition to improve the multiregional power source structure. The last step is to develop the electric power plan that includes a power source plan and a cross-regional power transmission plan.

(2) Making initial plans on the operation of the electric power system and the consumption of wind power and other renewable energy. First of all, it should be confirmed whether a quota system can be used as the constraint condition of renewable energy consumption. The introduction of a quota system may lead to changes in the renewable energy consumption plan and system operation plan. Initial plans on the operation of cross-provincial and cross-regional tie lines and on the consumption of wind and solar power should be drawn up according to the different renewable energy policy constraints.

(3) Improving the overall plan and scheduling. Improvements should be made in this regard according to the seasonal features, guaranteed capacity, effective output, and other parameters of hydropower and wind power, based on weekly or monthly electricity demand projections, and overall cycles of different

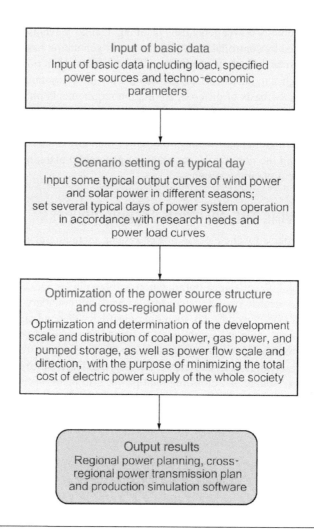

FIG. 3.3

Multiregional power source optimization planning process.

power sources, and for the purpose of reducing electricity loss, caused by inefficient use of water and wind, as much as possible.

(4) Making system production simulation analyses and optimizing the mode of operation and the non-fossil energy power generation consumption plan in accordance with system operation economic indicators. The production simulation and analysis process of the power system is shown in Fig. 3.4. In accordance with the power system plans and system production simulation, the total cost of power system operation (including fuel costs, power transmission costs, and external environmental expenses), can be worked out, and the operation mode of cross-provincial, cross-regional power transmission corridors shall be constantly adjusted in the process of production simulation

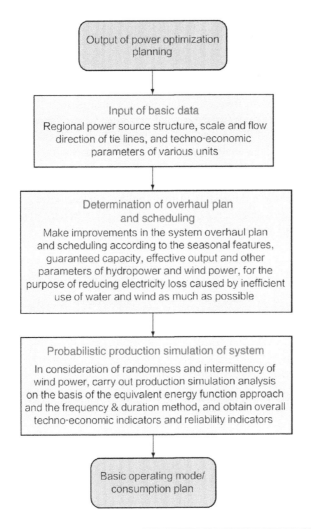

FIG. 3.4

The production simulation and analysis process of the electric system.

(see Appendix IV for the principles). If the difference between the renewable energy consumption and the total cost of system operation in either iterative computation is less than a certain set value, it shall be considered as calculation convergence, and the analysis results, including operation mode and renewable energy consumption plan, shall be output. Otherwise, the adjusted maintenance scheduling and unit commitment shall be returned.

(5) Analyzing system frequency stability and frequency control and verifying a renewable energy consumption scale and wind power curtailment. Carrying out system time-domain simulation at special times of difficult peak shaving, rapid load change and fast changes of wind speed and lighting, giving priority to frequency stability and frequency regulation results, analyzing the abundance of system frequency regulation capability and frequency stability, determining the period where the system frequency exceeds the limits, and giving the curtailed installed capacity of wind power and solar power as well as the estimated values of corresponding wind and solar power curtailments for maintaining system frequency stability in typical periods,[1] and amending the renewable energy consumption wind and solar power curtailments. The analysis process for frequency regulation verification is shown in Fig. 3.5, and the principle of the analysis model is shown in Appendix V.

(6) Obtaining China's consumption and proportion of non-fossil energy sources on the basis of the non-fossil energy consumption of regional grids by the target year. Working out the results of hydropower, nuclear power, wind power, biomass power, and solar power energy as well as water/wind/solar power curtailments for consumption through production simulation and frequency

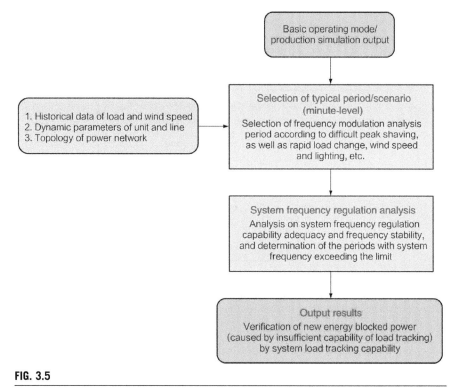

FIG. 3.5

The analysis process of frequency regulation verification.

[1]Wind and solar power curtailments refer to electricity that wind power farm and photovoltaic (PV) power station are capable of generating but fail to generate due to the factors of the capabilities of system balance adjustment and power grid transmission

regulation analysis. On the basis of the consumption of non-fossil energy power generated from regional grids, China's total non-fossil energy power consumption can be obtained by the target year. Moreover, by considering the total non-fossil energy consumption for purposes other than power generation, the proportion of non-fossil energy to primary energy consumption can be obtained to demonstrate the possibility of achieving the goal of 20% non-fossil energy use accounting for total energy use.

3.3 MODEL FOR EVALUATION OF ENERGY AND POWER DEVELOPMENT

3.3.1 CGE MODEL

The main feature of the CGE model (see Appendix VI for its theoretical basis), which is established on the basis of the general equilibrium theory, is that the model may help to make an analysis on the comprehensive results of all economic entities in market interactions, and study the interaction and interdependence between different elements. Regardless of external impact or policy change, as long as it affects the decision of an economic entity on supply or demand, the model will reflect the supply and demand changes of other economic entities, and spread the effects of the changes to the entire economy. Meanwhile, the CGE model is a multisector model that can quantitatively measure the direct effects of different non-fossil energy development paths on energy and power sectors, and its indirect effects on other industries. Therefore, compared with other models, it is a more detailed and comprehensive model for scenario evaluation.

Currently, the CGE model has been widely used to evaluate the policies on energy price, environmental taxation, trade, etc., and can be used as a tool for predicting various energy demands in the future. Based on the latest IO table (by province) of the National Bureau of Statistics of China, a multiregional dynamic CGE model is built to measure the impacts of electric power investments on China's regional energy, environment, and economy under different non-fossil energy development scenarios, such as the benefits of energy saving and environmental protection. The analysis process of the model is shown in Fig. 3.6.

The main characteristics of the multiregional dynamic CGE model established by the State Grid Energy Research Institute (SGERI) are shown as follows:

(1) The production and supply of the power sector have been refined. At present, there is only an "electric power, thermal production, and supply industry" for the power sector in the IO table, important basic data of the model. Based on the research data in the IO table prepared by the National Bureau of Statistics of China, the "electric power, thermal production, and supply sector" is further classified into thermal power, hydropower, nuclear power, and power generation based on other energy sources, electric power supply, as well as thermal production and supply. In this way, it can better reflect power

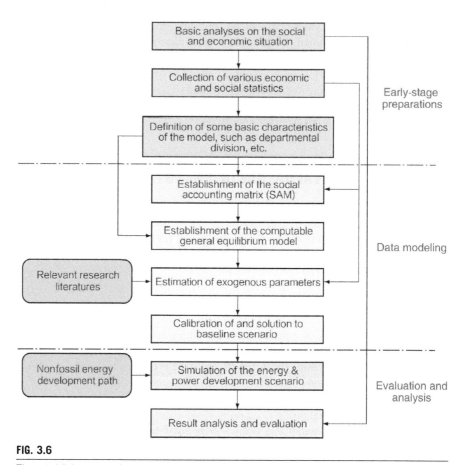

FIG. 3.6
The establishment and analysis process of the computable general equilibrium (CGE) model.

generation and the supply of non-fossil energy and fossil energy, detail the channels between different power suppliers and other economic activities, and simulate the impact of different non-fossil energy development paths on coordinated regional development.

(2) The differences between the investment in the power sector and that in other industries are described in detail. In the general multiregional CGE model, it is impossible to differentiate the investment composition since all investments are described in a unified way. Specifically, there are big differences between the investment in power source and those of other industries. For example, the proportion of construction investment in the electric power industry is smaller than in other industries, while the proportion of equipment investment in the electric power industry is larger than in other industries. In the model, the difference in the investment structure will have a direct impact on

the driving effects of power investment of other industries. Therefore, in the multiregion CGE model that is established by the SGERI, power investment is differentiated from other investments, and separate descriptions of different investment expenditure structures and capital accumulation can better reflect the driving effects of non-fossil energy development on the economy.

Here, we will focus on how to introduce power investments in the model.

(1) Power investments.

In order to improve the simulation, this model separates the investment in the power sector and that of other industries of the national economy, that is, different structures are adopted for the investment in the power sector and in other industries. Therefore, on the one hand, it can reflect the difference between investment in the power sector and in other industries; on the other hand, it provides the formation of the capital of the power sector in the dynamic part of the model with more accurate descriptions.

$$XAi_{i,r}^{ele} = \alpha_{i,r}^{ele} XI_r^{ele}$$

wherein XI_r^{ele} and $XAi_{i,r}^{ele}$ represent electric power investment and the demands for various investment goods in different regions and $\alpha_{i,r}^{ele}$ represents the share coefficient of electric power investment goods. Specifically, the setting of the model for $\alpha_{i,r}^{ele}$ is different. From the angle of simulation, the model will exogenously set the value in the electric power investment in accordance with the data of electric power investments.

(2) Capital formation of the power sector.

For the capital market, the model segments the entire capital market into two independent markets, namely the capital market (production capital or production capacity) of the power industry and the capital market of other industries. The model assumes that capital flow is not allowed between the two capital markets. For the capital market of other industries, the current capital stock is equal to the capital stock of the previous period, minus depreciation, plus the investments in the previous period. For the capital market of the power sector, the more accurate description is the fixed capital in the previous period, minus depreciation, and plus the newly added production capacity (power investment that has been completed in the previous period) in the previous period. The computational formula is as follows:

$$K_{\text{stock}_{ele,r}^t} = K_{\text{stock}_{ele,r}^{t-1}} (1 - \delta_{ele,r}) + XF_{ele,r}^{t-1}$$

wherein $K_{\text{stock}_{ele,r}^t}$, $K_{\text{stock}_{ele,r}^{t-1}}$, $\delta_{ele,r}$, and $XF_{ele,r}^{t-1}$ represent the fixed capital of the power sector of the regions in this period, the fixed capital of the power sector in the previous period, the depreciation rate of the power sector, and the investments in the power sector that have been completed in the previous period.

The differentiation between the electric power capital market and other capital markets may help to describe the capital flow between other industries.

For the capital flow of other industries, the constant elasticity of transformation (CET) capital supply equation below is used:

$$TR_r = \left[\sum_i \alpha_{i,r}^{ks} R_{i,r}^{(1+\sigma^{ks})} \right]^{\frac{1}{1+\sigma^{ks}}}$$

$$K_{i,r}^{'ks} = \alpha_{i,r}^{ks} = TK_r^s \left(\frac{R_{i,r}}{TR_r} \right)^{\sigma^{ks}}$$

wherein $R_{i, r}$, $K_{i, r}^{ks}$, and TK_r^s represent capital returns, capital supply, and total capital supply of other industries, respectively, σ^{ks} represents capital conversion elasticity. In other words, capital flow depends on the difference between and conversion elasticity of the intersector ratio of returns.

Besides electric power investment, power dispatching is also the main channel that influences the scenario simulation results. The overall power balance relational expression of the model is

$$XP_{c-ele}^r + \sum XP_{o-ele}^r + XDM_{c-ele}^r = XA_{ele}^r + XDE_{c-ele}^r$$

wherein XP_{c-ele}^r, $\sum XP_{o-ele}^r$, XDM_{c-ele}^r, XA_{ele}^r, and XDE_{c-ele}^r represent the output of coal power, the output of electricity generated from other power sources, input of coal power, total power demand of the region, and coal power delivered out of a certain region.

The left side of the equation is the total power supply of a certain region, namely electricity generated from coal and other power sources and power dispatched from other regions; while the right side of the equation is the total power demand of a certain region, namely regional power demand and the demand for power transmission out of the region.

For different areas, the electric power balance relation can also be expressed as

$$\begin{cases} XP_{c-ele}^r = XA_{ele}^r - \sum XP_{o-ele}^r - \overline{XDM_{c-ele}^r}, \\ XP_{c-ele}^z = XA_{ele}^z - \sum XP_{o-ele}^z + \overline{XDE_{c-ele}^z}, \end{cases}$$

where r represents the receiving region of coal power and z represents the sending region of coal power.

The first equation reflects the coal-fired power production in the coal receiving region. In the new model, the newly added power demands can mainly be met by construction of power plants, that is, for a coal power receiving region, on the basis of defining coal power dispatching (receiving of electricity) plan, power demands can be met by building power plants locally; while in a coal power sending region, electricity is mainly used for meeting local power demands, and is seldom delivered out of the region.

(3) It describes in details the costs of energy transportation and circulation. At present, China's multiregional CGE model generally does not define transportation and circulation costs. Considering that different non-fossil energy development paths have a big impact on the transportation

of power coal, this book defines the transportation and circulation costs of various energy communities at the time of establishing the model.

(4) It describes in details the mechanism of the formation of pollutants. In accordance with the mechanism, this model divides the emission of pollutants, such as sulfur dioxide and nitrogen oxide, as well as greenhouse gases including carbon dioxide, into three sources, namely pollutant emissions caused by energy as an intermediate fuel, departmental production processes, and final consumption, so as to better describe the impacts of the substitution effects caused by the changes of energy-use costs on the emission of pollutants and greenhouse gases in different paths of non-fossil energy development. The computational formula is shown as follows:

$$POL_{p,r} = \sum_i \left[emic(i,p) \sum_h XAC_{i,h,r} + emio(i,p)XP_{i,r} \right.$$
$$\left. + e\min(p,i,j) \sum_j XAP_{i,j,r} \right]$$

wherein $emic(i,p)$, $emio(i,p)$, and $e\min(p,i,j)$ represent the emission factors of various pollutants and carbon dioxide in the final consumption, production, and intermediate input processes, respectively, $POL_{p,\,r}$ is total emission of pollutants and carbon dioxide, $XAC_{i,\,h,\,r}$ is final consumer goods, $XP_{i,\,r}$ is departmental output, and $XAP_{i,\,j,\,r}$ is intermediate input.

3.3.2 THE IO MODEL

Power grid is an important infrastructure that concerns China's stability and people's well-being in the country, and the development of grid is directly or indirectly linked with other sectors of China's national economy, on an extensive scale and at multiple levels. The IO model may help to give a quantitative analysis on the influences of investment in power construction on China's economic growth, social employment, and the output of relevant sectors.

The IO model is a method for comprehensive survey and analysis on the quantitative interdependency between different sectors of the national economy and the links of social reproduction. It involves not only the direct driving effects of investment on the economy, but also the indirect effects of investment on all sectors of society by industrial association. To use the IO model to quantitatively measure the investment influence, first of all it is necessary to transform the fixed-asset investments in the grid into the demands of production sectors of relevant investment goods.

3.3.2.1 Analysis on grid investment flow

Measure the investment flows in accordance with power transformation and power transmission. The costs of power transformation cover construction costs, equipment procurement costs, installation costs, and other expenses, and the total cost of power transmission includes that of main materials and other expenses.

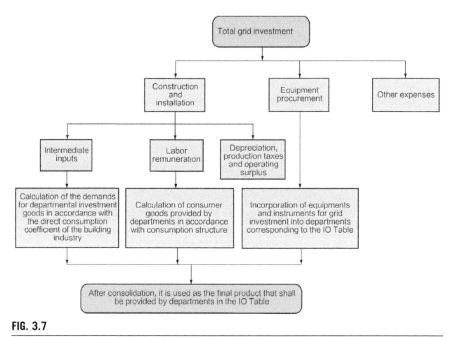

FIG. 3.7

The process of breakdown of grid investments.

Breakdown the grid investment flow according to the distribution ratio between expenses of power transformation and power transmission, and obtain construction and installation expenses of power transformation and transmission, equipment procurement costs, and other expenses in the grid investment. The breakdown of the grid investment is shown in Fig. 3.7.

3.3.2.2 Measurement of the impact of grid investment on employment and industrial outputs

The basic relational expression of the IO model is $X = (I - A)^{-1}Y$, wherein, $A = (a_{ij})_{n \times n}$ is the direct consumption coefficient matrix, $Y = (y_i)_{n \times 1}$ is the column vector of the final demand of the sector, and $X = (x_i)_{n \times 1}$ is the column vector of total output of the sector.

By adding up the above demands of "construction and installation costs" for various construction materials, the demands of "equipment procurement costs" for various equipment and instruments, and the demands of laborers for consumer goods, the total new demands of the total grid investment for the final products of all sectors (ΔY) can be obtained. By using the complete consumption coefficient matrix $(I - A)^{-1}$ in the IO table and the formula $\Delta X = (I - A)^{-1} \Delta Y$, the increment of the total new output of each sector for meeting the needs of power grid construction (ΔX) can be worked out; by multiplying ΔX and the added value rate of each sector (calculated from the IO table), the newly added value of each sector, as a result of grid investment can be worked out, and the increment of the gross domestic product (GDP) as a result of grid investment can be obtained by adding up the new added value of each sector; by working out the number of employees needed for the total unit output of each sector on the basis of *China Labor Statistical Yearbook*, and multiplying the increment of the above-calculated output of each sector (ΔX) by the number of employees for the unit output, the added quantity of employment of each sector driven by grid investment can be worked out.

3.3.3 ENERGY DEVELOPMENT QUALITY EVALUATION INDICATOR SYSTEM

Generally, the energy development quality evaluation indicator system consists of energy infrastructure data, analytical indicators, and evaluation indicators, see Fig. 3.8 for details. By building development scenarios, it may help to determine energy infrastructure, calculate analytical indicators, and conduct a comprehensive evaluation on energy development quality from the perspective of coordinated and sustainable development.

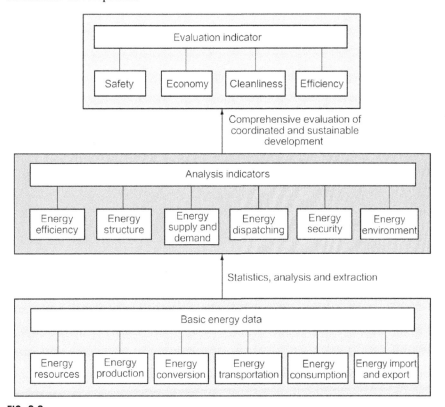

FIG. 3.8

The energy development quality evaluation indicator system.

See Appendix VII for the specific meaning of each indicator.

The energy infrastructure data cover energy resources, energy production, energy conversion, energy transportation, energy consumption, and energy import and export. Specifically, historical data are derived from various statistical yearbooks, and the data of future development are derived from the output results of the above model for integrated energy balance and optimization. Based on statistics, analysis, and extraction of energy infrastructure, analytical indicators are used to describe the characteristics of energy development, and are divided into six categories of energy efficiency, energy structure, energy supply and demand, energy dispatching, energy security, and energy environment. Evaluation indicators are established from the aspects of safety, economy, cleanliness, and efficiency, aiming to make a visible evaluation on energy development by establishing the overall indicators for comprehensive, integrated, and sustainable development.

Conditions and development potential of energy resources

4

Energy resources are the basis of energy development. China has abundant non-fossil energy resources with a potential for large-scale development. This chapter focuses on the resources of hydroenergy, nuclear energy, wind energy, solar energy, and biomass energy and their distribution, development situation, development potential, and relevant national plans. Meanwhile, it describes briefly fossil energy resources of coal, oil, and natural gas and their development potential.

4.1 NON-FOSSIL ENERGY

4.1.1 HYDROENERGY

4.1.1.1 Resources and their distribution

China is endowed with rich hydropower resources and ranks the first in the world in this aspect. According to China's hydropower resources review results released by the National Development and Reform Commission in 2005, there are a total of 3886 rivers, each of which has a theoretical reserve of water resources of 10,000 kW and above, in the Chinese mainland, with an annual energy output of 6.08 trillion kWh based on the total theoretical reserves and with an average power of 694 GW. China boasts an installed capacity of technically recoverable hydropower resources of 542 GW, and an annual power generation of 2.47 trillion kWh, as well as an installed capacity of economically recoverable hydropower resources of 402 GW, and an annual power generation of 1.75 trillion kWh. The hydropower resources of Chinese provinces are shown in Table 4.1.

China's hydropower resources mainly have the following features:

(1) Hydropower resources are rich in western China and scarce in eastern China. China is a country with a vast territory, and its terrain and precipitations vary greatly in different regions. Therefore, the hydropower resources of China feature an imbalanced geographical distribution, rich in the west and scarce in the east.[1] By the statistics of the installed capacity of technically recoverable

---

[1] The western region includes 12 provinces (autonomous regions and municipalities) including Yunnan, Guizhou, Sichuan, Chongqing, Shaanxi, Gansu, Ningxia, Qinghai, Xinjiang, Xizang, Guangxi, and Nei Mongol; the eastern region includes 10 provinces (municipalities), namely Beijing, Tianjin, Hebei, Shandong, Jiangsu, Zhejiang, Shanghai, Guangdong, Fujian, and Hainan.

Non-Fossil Energy Development in China. https://doi.org/10.1016/B978-0-12-813106-0.00004-0

Table 4.1 China's hydropower resources (by province, autonomous region, and municipality)

No.	Province (autonomous region and municipality)	Theoretical reserves		Technically available hydroenergy resources[a] (10 MW)	Economically available hydroenergy resources[b] (10 MW)	Installed capacity[c] at the end of 2012 (10 MW)
		Annual power generation (100 GWh)	Average power (10 MW)			
1	Beijing, Tianjin, and Hebei	199	227	175	125	285
2	Shanxi	444	563	402	397	243
3	Nei Mongol	509	581	262	257	111
4	Liaoning	178	203	177	173	287
5	Jilin	301	344	512	504	442
6	Heilongjiang	664	758	816	723	97
7	Shanghai and Jiangsu	152	174	6	2	114
8	Zhejiang	538	614	664	661	984
9	Anhui	274	312	107	100	278
10	Fujian	941	1074	998	970	1138
11	Jiangxi	426	486	516	416	418
12	Shandong	102	117	6	5	108
13	Henan	412	471	288	273	395
14	Hubei	1507	1720	3554	3536	3595
15	Hunan	1163	1327	1202	1135	1372
16	Guangdong	532	607	540	488	1303
17	Hainan	74	84	76	71	1513
18	Guangxi	1545	1764	1891	1858	82
19	Sichuan	12,572	14,351	12,004	10,327	609

20	Chongqing	2012	2296	981	820	3932
21	Guizhou	1584	1809	1949	1898	1728
22	Yunnan	9144	10,439	10,194	9795	3264
23	Xizang	17,639	20,136	11,000	835	54
24	Shaanxi	1119	1277	662	650	250
25	Gansu	1304	1489	1063	901	730
26	Qinghai	1916	2187	2314	1548	1101
27	Ningxia	184	210	146	146	43
28	Xinjiang	3344	3818	1656	1567	44
Total		60,829	69,440	54,163	40,180	24,890

Note: The above data come from the results of China's national hydropower resources review conducted in 2005.

[a]Technically available hydroenergy resources refer to hydroenergy resources that can be developed and utilized in theoretical hydroenergy resources and under present technical conditions.

[b]Economically available hydroenergy resources refer to, under the present technical and economic conditions, conventional hydropower resources that have higher economic development value but have no restrictive ecological and environmental problems in comparison with other power sources.

[c]Including pumped storage power station.

water resources, the hydropower resources in 12 provinces including Yunnan, Guizhou, Sichuan, Chongqing, Shaanxi, Gansu, Ningxia, Qinghai, Xinjiang, Xizang, Guangxi, and Nei Mongol, where the economy is relatively backward, account for 81.5% of the total amount of the country. Particularly, the hydropower resources in the five provinces in Yunnan, Guizhou, Sichuan, Chongqing, and Xizang in southwest China account for 66.7% of the total amount of the country; the hydropower resources in eight provinces including Heilongjiang, Jilin, Shanxi, Henan, Hubei, Hunan, Anhui, and Jiangxi account for 13.7% of the total amount of the country; while in the 10 provinces (municipalities), including Beijing, Tianjin, Hebei, Shandong, Jiangsu, Zhejiang, Shanghai, Guangdong, Fujian, and Hainan, which are economically developed and have concentrated electric loads, they account for a mere 4.6%.

(2) Cascade development is needed for runoff regulation due to the uneven temporal distribution of hydropower resources.

Situated in the southeast of the Eurasian continent, China borders the Pacific, the largest ocean in the world, and has a distinct monsoon climate. Therefore, the annual runoff and interannual runoff distributions of most rivers are uneven. There is a wide difference between flow in the flood season and in the dry season. The water volume of China's major rivers, during the flood season, accounts for 70%–80% of the total runoff of the year, and, during the dry season, accounts merely for 20%–30% of the total runoff of the year. Taking the mainstream of the Yangtze River as an example, it has rich hydropower resources but the annual distribution of the precipitation it receives is very uneven, and concentrates mainly in May and October in the flood season, accounting for 70%–90% of the annual precipitation. Similarly, the annual distribution of the runoff volume concentrates mainly in the flood season, especially in June and August. For this, it is necessary to build reservoirs with a good regulation performance, so as to regulate the runoff, raise the overall quality of hydropower generation, better meet the requirements on the power market, and give play to the comprehensive benefits of flood control and shipping, etc.

(3) China's hydropower resources are mainly distributed in main streams of big rivers and are suitable for large-scale and centralized development.

Geographically, China's hydropower resources concentrate in big rivers including the Yangtze River, the Yellow River, the Jinsha River, the Yalong River, and the Dadu River, and the Chinese government has planned 13 large hydropower bases (hereinafter referred to as the 13 hydropower bases), namely the upper stream of the Yangtze River, the Jinsha River, the Dadu River, the Yalong River, the Wujiang River, the Nanpan River, and Hongshui River, the main stream of the Lancang River, the upper reach of the Yellow River, the main stream of the Yellow River, northeast China, western Hunan, Minzhegan (Fujian, Zhejiang, and Jiangxi), and the Nujiang River. The total installed capacity of technically recoverable resources of the 13 hydropower bases is 280 GW, accounting for about 52% of the total technically recoverable hydropower resources of China. Particularly, the technically recoverable

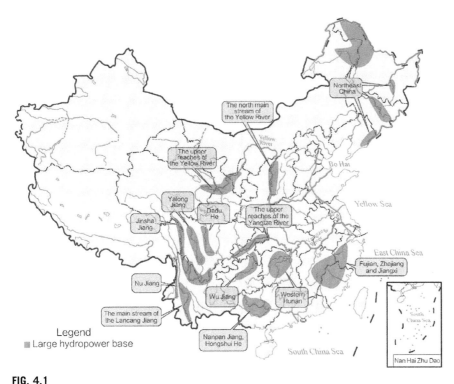

FIG. 4.1

A schematic map of China's 13 major hydropower bases.

hydropower resources of the Jinsha River, the Yalong River, the Dadu River, the Nu River, and the main stream of the Lancang River accounts for 56.9% of that of the 13 hydropower bases. The technically available hydroenergy resources of the main stream of the middle and lower reaches of the Jinsha River is 58.58 GW, that of the upper stream of the main stream of the Yangtze River (from Yibin to Yichang) is 30.15 GW, that of the Yalong River, the Dadu River, the upper reach of the Yellow River, the Lancang River, and the Nu River has exceed 20 GW, and that of the Wu River as well as the Nanpan River and Hongshui River has exceeded 10 GW. The above rivers are rich in hydropower resources, have favorable conditions for cascade and rolling hydropower development, building large hydropower bases and fully leveraging the scale benefits of hydropower resources. The distribution of the 13 hydropower bases is shown in Fig. 4.1 (Table 4.2).

4.1.1.2 Development situation

During the initial period of its founding, China undertook hydropower construction activities mainly in its eastern region where the economic development and power utilization grew fast, but built few large hydropower stations. At the end of the 1950s, it began to build large hydropower stations including the Liujiaxia Hydropower

Table 4.2 Technically available hydroenergy resources of the 13 hydropower bases (unit: MW)

Hydropower base	Technically available hydroenergy resources
Northeast China	18,690
The north main stream of the Yellow River	6408
The upper reaches of the Yellow River	20,032
Yalong River	25,310
Dadu River	24,596
The upper reaches of the Yangtze River	33,197
Jinsha River	58,580
Fujian, Zhejiang, and Jiangxi	10,925
Nu River	21,420
Wu River	10,795
Western Hunan	5992
The main stream of the Lancang River	25,665
Nanpan River and Hongshui River	14,313
The 13 hydropower bases (total)	275,773

Note: The above data are the results of China's hydropower resources review conducted in 2005.

Station on the main stream of the Yellow River, although it still gave priority to the development and construction of large hydropower stations in eastern China. Since reform and opening up of the country, China has given greater priority to the development of the hydropower resources in its western region. Especially, after the "West-to-East Power Transmission" strategy was put forward, the abundant hydropower resources in southwest China have been greatly developed and gradually utilized. Besides the Three Gorges Hydropower Station on the main stream of the Yangtze River, large hydropower stations have been built or are being built on the important tributaries of the Yangtze River, including the Yalong River, the Dadu River, and the Wu River, and a batch of key hydropower stations with an installed capacity of over 1 GW have been built or being built on the reaches of the Yellow River between the Longyang Gorge and the Qingtong Gorge, the Lancang River, the Nanpan River, and the Hongshui River. The development of hydropower resources on the rivers or reaches has laid a foundation for the large-scale development of the western region and the successful implementation of the "West-to-East Power Transmission" project. Statistical data show that, in 1980 at the initial stage of China's reform and opening-up, China's total installed power capacity was 65.87 GW, including an installed hydropower capacity of 20.32 GW, accounting for 30.8% of the total. The installed hydropower capacity increased by about 16 GW during 1980–90 and about 43 GW between 1990 and 2000. Since the 21st century, China has entered a period of large-scale and rapid development, and its installed hydropower capacity increased by about 230 GW between 2001 and 2015.

In 2015, China's installed hydropower capacity surpassed 310 GW and hit 319 GW, of which the pumped storage power station reached 22.71 GW. In 2015, the net increase of China's installed hydropower capacity was 17.54 GW, mainly distributed in central China and southern China, with 7.19 million and 8.01 GW, accounting for 41.7% and 45.6% of China's net increase of installed hydropower capacity, respectively. As is seen from the distribution of operating hydropower stations, central China and southern China have the highest installed hydropower capacity in the country, with an installed capacity of 137 and 109 GW at the end of 2015, accounting for 42.8% and 34.1% of the total, respectively. The regional distribution of China's installed hydropower capacities at the end of 2014 and at the end of 2015 is shown in Fig. 4.2.

By province, Sichuan, Yunnan, and Hubei have the largest installed hydropower capacity in China, exceeding 30 GW and reaching 69.39, 57.74, and 36.53 GW (including pumped storage power stations), respectively. The installed hydropower capacity exceeded 10 GW in other provinces, such as Guizhou, Guangxi, Hunan, Guangdong, Fujian, Qinghai, and Zhejiang, with relatively abundant hydropower resources. The installed hydropower capacity (including pumped storage power stations) of Chinese provinces in 2015 is shown in Fig. 4.3.

As is seen from the development of pumped storage power stations, as at the end of 2014, China had built 26 pumped storage power stations, with an installed capacity of 22.72 GW; it has 17 pumped storage power stations under construction, with an

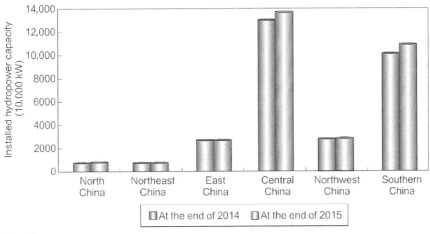

FIG. 4.2

Regional distribution of China's installed hydropower capacity at the end of 2014 and at the end of 2015. Note: Northeast China includes Liaoning, Jilin, Heilongjiang, and eastern Nei Mongol; North China includes Beijing, Tianjin, Hebei, Shanxi, Shandong, and Western Nei Mongol; East China includes Shanghai, Jiangsu. Zhejiang, Anhui, and Fujian; Central China includes Henan, Hubei, Hunan, Jiangxi, Sichuan, and Chongqing; Northwest China includes Shaanxi, Gansu, Qinghai, Ningxia, and Xinjiang; Southern China includes Guangdong, Guangxi, Yunnan, Guizhou, and Hainan.

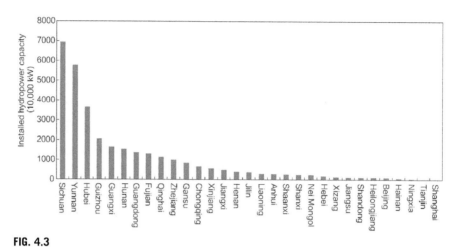

FIG. 4.3

The installed hydropower capacities of Chinese provinces in 2015.

installed capacity of 21.14 GW. The built and under-construction pumped storage power stations are basically distributed in load center regions including North China, East China, Central China, and Southern China Grid, accounting for about 90%. Built and under-construction pumped storage power stations of China's regional grids in 2014 are shown in Fig. 4.4.

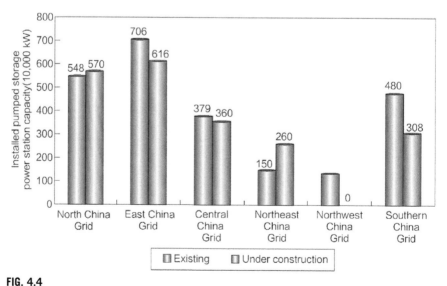

FIG. 4.4

Built and under-construction pumped storage power stations of China's regional grids in 2014.

4.1.1.3 Development potential

As is seen from the development of the 13 hydropower bases, the eight hydropower bases of the upper reach of the Yangtze River, the Wu River, the Nanpan River, and the Hongshui River, the upper reach and the northern main stream of the Yellow River, Western Hunan, Minzhegan (Fujian, Zhejiang, and Jiangxi), and northeast China hydropower base have been highly developed; the five hydropower bases of the Jinsha River, the Yalong River, the Dadu River, the Lancang River, and the Nu River are less developed, all of them being in southwest China. In accordance with the latest statistics, there are a total of 113 hydropower stations undergoing planning on the main streams of the above five hydropower bases, with a planned total installed capacity of 263 GW. At present, the installed capacity of operating hydropower stations on the main streams of the hydropower bases in southeast China is small, with a development rate of less than 10% and great development potential.

China has relatively abundant resources of pumped storage power station sites, and the site resources, for which surveys have been made have a large total capacity. According to the construction conditions and the early site selection planning of pumped storage power stations of the regional grids, the total capacity of early site selection for the pumped storage power stations, within the regions operated by the State Grid Corporation of China, may reach 130 GW, see Table 4.3 for details.

4.1.1.4 Development planning

In accordance with the *Strategic Action Plan on Energy Development (2014–2020)* released by the General Office of the State Council, as well as early-stage construction of hydropower projects and the relevant results of grid planning, it is estimated that the total installed hydropower capacity will reach 380 GW in 2020. Specifically, the installed capacity of conventional hydropower will reach 350 GW, and that of pumped storage will reach 35 GW.

Table 4.3 Total capacity of early-selected pumped storage power stations in the areas operated by the State Grid Corporation of China (unit: 10,000 kW)

Early-selected site region	Number of sites	Total capacity
Regions operated by the State Grid Corporation of China (total)	114	13,236
North China	24	3100
East China	28	3615
Central China	28	3301
Northeast China	17	1760
Northwest China	17	1460
Xizang	0	0

As is seen from hydropower distribution, China will make intensive efforts to build large hydropower energy bases at the lower reach of the Jinsha River, the Yalong River, the Dadu River, and the Lancang River in its western regions, accelerate the development of the hydropower bases at the middle reach of the Jinsha River, start the construction of the large hydropower energy bases at the upper reach of the Jinsha River and the middle and lower reaches of the Nu River, and carry out the construction of pumped storage power stations in an orderly manner. In its central region, China will reasonably develop the remaining hydroenergy resources of the main stream of the Yellow River, the lower reach of the Han River, the Lou River, the Du River, and the Gan River, and speedup the construction of pumped storage power stations in an appropriate manner. In its eastern region, China will give priority to expansion, renovation, and upgrading of existing power stations, and strengthen the construction of pumped storage power stations.

Although China has a resource potential and technological advantages for large-scale hydropower development, it faces problems and challenges such as land inundation, eco-environmental protection, resettlement, development difficulty and international river development, etc. Therefore, overall planning shall be made to ensure harmonious development. In the process of hydropower development, the Chinese government shall attach great importance to ecological protection, strengthen environmental impact assessment of river hydropower planning, implement environmental protection measures for hydropower projects, and place equal emphasis on environmental protection and hydropower development. For the resettlement of affected residents in reservoir areas, we shall summarize relevant experience, develop and improve policies and measures for ensuring the social security of resettled residents, urbanized settlement, and construction, and explore and promote channels of resettlement.

4.1.2 NUCLEAR ENERGY

4.1.2.1 Resources and their distribution

(1)   Resource reserves.

So far, uranium resources that have been explored and proven are mainly located in Liaoning, Nei Mongol, Xinjiang, Jiangxi, Hunan, and Guangdong, and nearly 50% of uranium ores have not been explored, indicating that there is a great potential reserve and a promising future.

So far, China has proven over 300 uranium mines of different sizes and confirmed considerable uranium reserves. The distribution of uranium resources is shown in Fig. 4.5. According to the data released by the Organization for Economic Cooperation and Development (OECD) in 2003, China has 77,000 tons of uranium (at a cost of less than USD 130/kg uranium), of which the reserves of the uranium resources, at a cost of less than USD 40/kg, account for approximately 60% of the total uranium resources of the country. In China, most uranium ores are medium or low grade, the uranium ores with a grade of 0.05%–0.3% account for the vast majority of the total resources, small

FIG. 4.5

A schematic map of China's uranium ore resources.

Data from: http:/www.mining120.com.

and medium-sized ore deposits take a large part of the total reserve (accounting for over 60% of the total reserve), and the burial depths of proven uranium deposits are mostly less than 500 m. Table 4.4 shows information on China's operating uranium mines.

At the beginning of 2008, China's first 10,000-ton leachable sandstone uranium deposit was discovered in Yili region, Xinjiang, marking a major breakthrough of exploration. Currently, China is carrying out a new round of prediction and evaluation of the potential of uranium resources, and its potential reserve is likely to reach several million tons.

(2) The impacts of technological advance on availability of uranium resources.

Uranium resource reserve is a natural concept, and, more importantly, a technical and economic concept. Economically available uranium resources vary with the price fluctuations of the international uranium market. As uranium ore mining technology advances, it is possible for uranium ores that were originally uneconomical to become economically recoverable. Generally, there is a considerable potential of natural uranium resources in the world. By far, all researchers are optimistic about uranium resource reserves for the

Table 4.4 China's uranium mines in operation

Mine	Type	Production capacity (ton natural uranium/year)	Starting time of mining (year)
Wuzhou, Jiangxi	Underground and open-pit mine	300	1966
Chongyi, Jiangxi	Underground and open-pit mine	120	1979
Yining, Xinjiang	In situ leaching (ISL)	200	1993
Lantian, Shaanxi	Underground	I	1993
Benxi, Liaoning	Underground	12	1996

Note: the above data come from the Nuclear Power in China, http:/www.world-nuclear.org.

development of nuclear power, and some even consider that the world's uranium resources are sufficient in the next 500 years.[2]

In the long run, if it is possible to build a fast-reactor nuclear power station with higher efficiency, its utilization ratio of nuclear fuel is expected to reach 60%–70%, rather than the approximate 1% at present, then the supply of nuclear fuel will no longer be a problem. In the mode of a fast reactor, the available quantity of China's proven uranium resources will be significantly improved.

4.1.2.2 Development situation

In 2015, the Fangjiashan Nuclear Power Plant No. 2 Unit, the Yangjiang Nuclear Power Plant No. 2 Unit, the Ningde Nuclear Power Plant No. 3 Unit, the Hongyanhe Nuclear Power Plant No. 3 Unit, the Fuqing Nuclear Power Plant No. 2 Unit, and the Changjiang Nuclear Power Plant No. 1 Unit were put into operation. As at the end of 2015, 28 nuclear power units had been put into commercial operation in China, with a total installed capacity of about 26.46 GW. Specifically, the installed capacity of nuclear power in Zhejiang totals 6.56 GW, including 0.31 GW of the Qinshan No. 1 Nuclear Power Plant, 2.61 GW of the Qinshan No. 2 Nuclear Power Plant, 1.46 GW of the Qinshan No. 3 Nuclear Power Plant, and 2.18 GW of the Fangjiashan Nuclear Power Plant. In Guangdong it totals 8.32 GW, including 1.97 GW of the Daya Bay Nuclear Power Plant, 1.98 GW of the Ling'ao Nuclear Power Plant, 2.19 GW of the Ling'ao No. 2 Nuclear Power Plant, and 2.18 kW of the Yangjiang Nuclear Power Plant. In Fujian it totals 5.45 GW, including 3.27 GW of the Ningde Nuclear Power Station and 2.18 GW of the Fuqing Nuclear Power Station. In the Tianwan Nuclear Power Station of Jiangsu it is 2.12 GW. In Liaoning in the Hongyanhe Nuclear Power Station (Phase 1) it is 3.36 GW; and in the Changjiang Nuclear Power Station of Hainan it is 0.65 GW. The changes of installed nuclear power capacity since the "12th Five-year Plan" period are shown in Table 4.5.

---

[2]Richard D. Wilson and Robert Krakowski, Planning for Future Energy Resources, Letter for Science, 2003 (300); 25.

Table 4.5 The changes of installed nuclear power capacities during the 12th "Five-year Plan" period (unit: 10 MW)

Year	2010	2011	2012	2013	2014	2015
Total	1088	1264	1264	1485	2033	2646
Zhejiang	373	438	438	438	547	656
Qinshan No. 1 Nuclear Power Station	31	31	31	31	31	31
Qinshan No. 2 Nuclear Power Station	196	261	261	261	261	261
Qinshan No. 3 Nuclear Power Station	146	146	146	146	146	146
Fangjiashan Nuclear Power Station					109	218
Guangdong	503	614	614	614	723	832
Dayawan Nuclear Power Station	197	197	197	197	197	197
Ling'ao Nuclear Power Station	198	198	198	198	198	198
Ling'ao No. 2 Nuclear Power Station	108	219	219	219	219	219
Yangjiang Nuclear Power Station					109	218
Fujian				109	327	545
Ningde Nuclear Power Station				109	218	327
Fuqing Nuclear Power Station					109	218
Liaoning				112	224	336
Hongyanhe Nuclear Power Station				112	224	336
Jiangsu	212	212	212	212	212	212
Tianwan Nuclear Power Station	212	212	212	212	212	212
Hainan						65
Changjiang Nuclear Power Station						65

At the end of 2015, China had the largest nuclear power capacity under construction in the world, with 26 nuclear power units being built and an installed capacity of 28.7 GW. Table 4.6 lists China's under-construction nuclear power projects at the end of 2015.

In addition, the site survey or preliminary work of nuclear power stations is being conducted in the inland provinces including Hubei, Hunan, Jiangxi, Sichuan, Henan, Gansu, and Nei Mongol. The distribution of China's existing, under-construction and planned nuclear power stations is shown in Fig. 4.6.

Table 4.6 China's under-construction nuclear power units at the end of 2015

No.	Name of unit	Capacity (10 MW)	Remarks
1	Liaoning Hongyanhe Nuclear Power Station, Phase 1	112	Started in August 2009, began operating in September 2016
2	Liaoning Hongyanhe Nuclear Power Station, Phase 2	2 × 112	Started in March 2015, to be operated in 2020
3	Shandong Haiyang Nuclear Power Station, Phase 1	2 × 125	Started in December 2009, the first unit to be operated in 2018
4	Huaneng Shidaowan Nuclear Power Station	20	Started in December 2012, to be operated in 2018
5	Zhejiang Sanmen Nuclear Power Station	2 × 125	Started in April 2009, to be operated in 2018
6	Jiangsu Tianwan Nuclear Power Station, Phase 2	2 × 100	Started in December 2012, to be operated in 2018
7	Jiangsu Tianwan Nuclear Power Station, Phase 3	100	Started in December 2015, to be operated in 2020
8	Fujian Ningde Nuclear Power Station, Phase 1	108	Started in September 2010, began operating in July 2016
9	Fujian Fuqing Nuclear Power Station	2 × 108	Started in December 2010, began operating in 2016 and 2017, respectively
10	Fujian Fuqing Nuclear Power Station, Phase 2	2 × 108	Started in May 2015, to be operated in 2020
11	Guangdong Taishan Nuclear Power Station, Phase 1	2 × 175	Started in December 2009, to be operated in 2018 and 2019
12	Guangdong Yangjiang Nuclear Power Station	4 × 108	Started in November 2010, two units began operating in 2016 and 2017, respectively, and the other units are to be operated during 2018–19
13	Guangxi Fangchenggang Nuclear Power Station, Phase 1	2 × 108	Started in July 2010, the No. 1 unit was began operating in January 2016, and the No. 2 unit was operated in October 2016
14	Guangxi Fangchenggang Nuclear Power Station, Phase 2	108	Started in December 2015, to be operated in 2020
15	Hainan Changjiang Nuclear Power Station	65	Started in November 2010, began operating in 2016

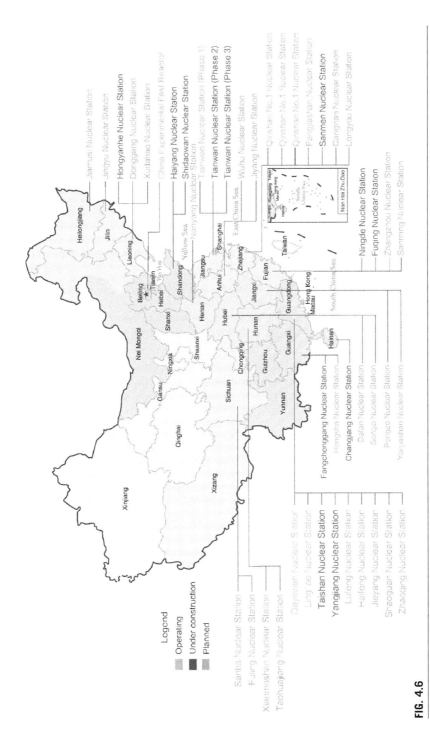

FIG. 4.6

The distribution of existing, under-construction and planned nuclear stations in China (as at the end of 2015).

4.1.2.3 Development potential

In recent years, by making further innovation on the basis of absorbing and introducing foreign advanced technologies, China has the basis and conditions for large-scale nuclear power development in terms of independent design, manufacturing, construction and operation of nuclear power, as well as site resource reserves, nuclear fuel supply, operation safety, and economical efficiency. In addition, although all operating Chinese nuclear power plants are located in coastal areas, there are no special restrictions on inland nuclear power construction in terms of technology and environmental protection, etc.

As is seen from the capability of nuclear fuel supply, China is a country with relatively abundant uranium resources and a promising future of mineral exploration. China has established a long-term friendly and cooperative relationship with major uranium-producing countries and uranium companies, and has done a lot of preliminary work in terms of its overseas exploration of uranium resources. Generally, by strengthening uranium resource exploration and development, strengthening international cooperation and making active use of overseas resources, China will not be restricted by the supply of natural uranium in its large-scale development of nuclear power.

As is seen from the reserves of site resources, according to the incomplete statistics, the installed capacity of China's resources in sites, where construction has started and the feasibility study review has been passed, is more than 70 GW; taking backup sites into consideration, China's current site resources may support an installed nuclear power capacity of 160 billion kW; by further site selection and exploration, China's nuclear power plant site resources are expected to meet the target of an installed capacity of 300–400 GW.[3]

4.1.2.4 Development planning

Since 2003, China's nuclear power has shifted gradually from "appropriate development" to "positive development." According to the *Medium and Long-term Development Plan on Nuclear Power (2005–2020)*, developed by the Chinese government in 2005, by 2020, China plans to build nuclear power units with a total installed capacity of 40 GW and ensures that nuclear power units with an installed capacity of 18 GW will be under construction at that time. According to the *Strategic Action Plan on Energy Development (2014–2020)*, by 2020, China plans to raise its installed nuclear power capacity to 58 GW, and raise its installed capacity of nuclear power stations under construction to over 30 GW. As is seen from the scale of nuclear power stations that have been built or are under construction, it is estimated that the installed nuclear power capacity will reach 54 GW by 2020.

It can be predicted that in the next 20 years, China's nuclear power will enter a period of rapid development. Although it may undergo twists and turns due to the

---

[3]National Energy Administration, Scientific Development in 2030: Research Report on China's National Energy Strategy.

Fukushima nuclear disaster in Japan, in the long run, China will pursue the policy on nuclear power unswervingly. On the basis of policy of "ensuring an efficient development of nuclear power on the basis of safety" defined in the "12th Five-year Plan" and on the principle of "Safety First, Quality First," the Chinese government follows the strictest standards, to carry out safety evaluations on all nuclear power facilities and strengthen the whole-process management of nuclear power design, construction, and operation, so as to eliminate all hidden dangers and ensure absolute safety.

4.1.3 **WIND ENERGY**

4.1.3.1 Resources and their distribution

In 2009, the China Meteorological Administration, in line with arrangements in the *Implementation Opinions on Promoting the Development of the Wind Power Industry*, issued by the National Development and Reform Commission and the Ministry of Finance, carried out a new round of general exploration and evaluation of wind energy resources. It established a professional network for observation of wind energy, consisting of 400 wind towers of 70, 100, and 120 m nationwide, and established a model for evaluation of China's wind energy resources. On the basis of the numerical simulation results of national wind energy resources with a resolution of 5 km × 5 km, if the areas (such as water body, wetland, marshland, nature reserve, historical site, and national park) with wind energy resources that are impossible to be developed are not considered, it has obtained the potential development capacity of wind energy resources at a level of two, three, and four, with a ground height of 50, 70, and 110 m, and has emphatically evaluated the wind energy resources in Nei Mongol (eastern Nei Mongol and western Nei Mongol), Hami (Xinjiang), Jiuquan (Gansu), the Bashang Plateau of Hebei, western Jilin, and Jiangsu.

The evaluation results show that, the potential development capacity of China's onshore wind energy resources (with a ground height of 50 m, at a level of at least three, with a wind power density of at least $300 W/m^2$) is about 2.38 billion kW (excluding the Qinghai-Xizang Plateau), while that of China's offshore wind energy resources (at the areas with a water depth of 5–25 m, a height of 50 m, and a level of at least three) is about 200 GW.

(1) Onshore wind energy resources.

The potential development capacity of China's onshore wind energy resources (at a level of at least four, with a wind power density of at least $400 W/m^2$) is 1.13–2.31 billion kW, that of onshore wind energy resources (at a level of at least three, with a wind power density of at least $300 W/m^2$) is 2.38–3.8 billion kW, that of onshore wind energy resources (at a level of at least two, with a wind power density of at least $200 W/m^2$) is 3.94–5.73 billion kW, see Table 4.7. The distribution of China's wind energy resources is shown in Fig. 4.7.

Table 4.7 The potential development capacity of China's onshore wind energy resources (unit: 100 GW)

Height above the ground (m)	Level 4 and above (wind power density not less than 400 W/m²)	Level 3 and above (wind power density not less than 300 W/m²)	Level 2 and above (wind power density not less than 200 W/m²)
50	11.3	23.8	39.4
70	15.1	28.5	47.9
110	23.1	38.0	57.3

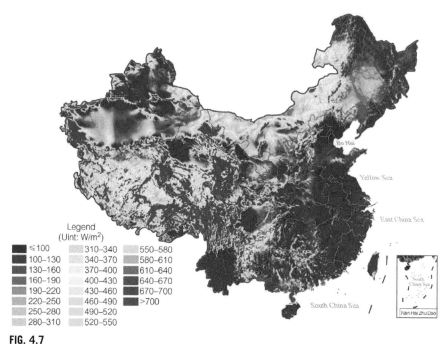

Legend
(Uint: W/m²)

≤100	310–340	550–580
100–130	340–370	580–610
130–160	370–400	610–640
160–190	400–430	640–670
190–220	430–460	670–700
220–250	460–490	>700
250–280	490–520	
280–310	520–550	

FIG. 4.7

A schematic map of China's wind energy resources.

From: Wind and Solar Energy Evaluation Center, China Meteorological Administration.

(2)   Offshore wind energy resources.

The evaluation results of wind energy resources show that, the coastal areas of Fujian and southeast Zhejiang near the Taiwan Strait are regions with the largest offshore wind energy resources, with a wind energy resource level of six or above; the coastal areas in northeast Zhejiang and Guangdong as well as the western coastal waters of the Hainan Island also have abundant wind energy resources, with a wind energy resource level of four to six; besides, other coastal areas of China (including Liaoning, Hebei, Shandong, Jiangsu,

FIG. 4.8

The distribution of annual average wind power density in China's offshore areas with a water depth of 5–25 m.

From: Wind and Solar Energy Evaluation Center, China Meteorological Administration.

and the Beibu Gulf of Guangxi) also have abundant resources, with a wind energy resource level of three to four.

Considering that the development of offshore wind energy resources is significantly influenced by the water depth, the technologies for offshore wind power development with a water depth of 5–25 m are relatively mature, while that with a water depth of 25–50 m still need to be improved. The evaluation results of wind energy resources show that, the potential development capacity of China's offshore wind energy resources with a water depth of 5–25 m is about 190 GW, and that with a water depth of 25–50 m is 21 GW. The distribution of annual average wind power density in China's offshore areas with a water depth of 5–25 m is shown in Fig. 4.8.

4.1.3.2 Key areas of wind energy development
As is seen from the distribution of China's wind energy resources, the large wind bases of eastern Nei Mongol, western Nei Mongol, Hami (Xinjiang), Jiuquan (Gansu), the Bashang Plateau of Hebei, Jilin, and the coastal areas of Shandong

Table 4.8 Evaluation of wind energy resources in key regions (with a height of 50 m, and with a level of at least three)

Name of base	Area of resources (10,000 km²)	Potential development capacity (10,000 kW)
Nei Mongol (eastern Nei Mongol and western Nei Mongol)	33.64	130,530
Hami (Xinjiang)	5.35	24,910
Jiuquan (Gansu)	4.70	20,520
Hebei (Bashang Plateau)	3.44	7930
Western Jilin	0.45	1540
Coastal areas of Shandong	–	3947
Coastal areas of Jiangsu (sea area with a water depth between 5 and 25 m)	4.62	1390
Key regions (total)	–	19,767

Note: *The above data come from the* China Energy Development Report 2010 *of the National Energy Administration, May 2010, and the* General Research Report on Grid Connection of Market Consumption of Wind Power *of the National Energy Administration, July 2011.*

and Jiangsu have abundant wind energy resources. The potential development capacity of wind energy resources with a height of 50 m and at a level of at least three is about 1.91 billion kW, accounting for about 80% of China's potential development capacity, see Table 4.8 for evaluation of wind energy resources in key regions.

4.1.3.3 Development situation

In recent years, driven by the *Renewable Energy Law* and supporting policies, China's wind power industry has grown rapidly and preliminarily achieved large-scale and industrialized development, and wind power has become an important part of China's new energy industry.

China's installed grid-connected wind power capacity and generated power between 2005 and 2015 are shown in Table 4.9. As of the end of 2015, the installed grid-connected wind power capacity had reached 128.3 GW, up by 32.5% on a year-on-year basis, accounting for 8.5% of China's total installed wind power capacity. During the "12th Five-year Plan" period, the annual average growth rate of installed grid-connected wind power capacity was 34.1%, and that of the wind power generation reached 30.2%, much higher than the growth rate of the total installed wind power capacity and the annual total power generation of China.

By province, in 2015, all of the installed grid-connected wind power capacities of the nine provinces including Nei Mongol, Xinjiang, Gansu, Hebei, Ningxia, Shandong, Shanxi, Liaoning, and Heilongjiang exceeded 5 GW, and the total installed grid-connected wind power capacity reached 97.45 GW, accounting for approximately 76% of the total installed grid-connected wind power capacity of the country. Specifically, at the end of 2007, Nei Mongol became the first

Table 4.9 Installed grid-connected wind power capacity and power generation

Year	Installed grid-connected wind power capacity (10,000 kW)	Proportion of installed wind power capacity (%)	Electricity generated from wind energy (100 GWh)	Proportion of electricity generated from wind energy (%)
2005	106	0.20	16	0.07
2006	207	0.33	28	0.09
2007	420	0.58	57	0.17
2008	839	1.06	131	008
2009	1760	2.01	276	005
2010	2958	306	494	117
2211	4623	4.35	771	1.57
2012	6083	501	1004	2.02
2013	7631	600	1393	2.60
2014	9686	700	1599	2.87
2015	12,830	802	1851	300

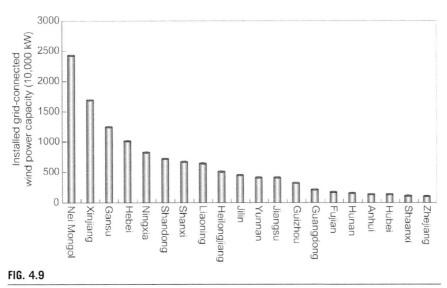

FIG. 4.9

Installed grid-connected wind power capacities of the main Chinese provinces in 2015.

province with an installed grid-connected wind power capacity of more than 1 GW; it continued to be far ahead of other provinces at the end of 2015, with its installed grid-connected wind power capacity reaching 24.25 GW. The installed grid-connected wind power capacities of the main Chinese provinces in 2015 are shown in Fig. 4.9.

Table 4.10 The estimated wind curtailments in some provinces (autonomous regions) in 2015

Province (autonomous region)	Statistics concerning installed wind power capacity (10,000 kW)	Power generation (100 GWh)	Wind curtailment (100 GWh)	Proportion of wind curtailment (%)
Gansu	1250	127	82	39
Jilin	444	60	27	32
Xinjiang	1611	148	70	32
Heilongjiang	503	72	19	21
Xinjiang Production and Construction Corps	80	4	1	19
Nei Mongol	2425	408	91	18
Ningxia	822	88	13	13
Liaoning	639	112	12	10
Hebei	1022	168	19	10
Yunnan	412	94	3	3
Shanxi	669	100	3	2
Total	12,830	1851	339	15

At present, there is a large installed wind power capacity in China's "Three Norths" (Northwest China, North China, and Northeast China) regions. However, due to the inadequate peak shaving capabilities of local power systems and the backward situation of cross-provincial, cross-regional power construction, the capability of local wind power consumption markets is insufficient, and there is a serious problem of wind curtailment in the regions. According to the relevant statistics of the National Energy Administration concerning wind power development in 2015, China's wind power curtailment was approximately 33.9 TWh during the year, with a proportion of wind power curtailment of 15%. Assuming that China's national average standard coal consumption for power supply was 315 g standard coal/(kWh) in 2015, the wind power curtailment was equivalent to about 10 million tons of standard coal. Specifically, the proportion of wind curtailment in Gansu, Nei Mongol, and Jilin exceeded 20%, and that in Heilongjiang and Liaoning exceeded 10%. The statistical data of wind curtailment of some provinces in 2015 is shown in Table 4.10.

4.1.3.4 Development potential

As is seen from the overall distribution of China's wind energy resources, the high-value areas are mainly located in the "Three Norths" regions and eastern coastal areas, where the wind energy resource reserve and technical available wind energy resources account for over 80% of the country's total, with great development potential of wind energy resources. From the perspective of the utilization of wind energy resources, the Chinese government shall focus on wind power development in these areas.

For the construction of wind power farms, besides considering wind energy resources, we need also to consider the factors of engineering geology, site, natural hazards, and land or offshore development and utilization. According to the preliminary results of China's wind power construction, the areas that are suitable for the construction of large-scale wind power bases are mainly Jiuquan of Gansu, Hami and Dabancheng of Xinjiang, northern Ningxia, Liaoning, western Jilin, Heilongjiang, eastern Nei Mongol, western Nei Mongol, northern Hebei, northern Shanxi, the coastal areas of Shandong, and the coastal areas of Jiangsu. The areas are located in the "Three Norths" area or coastal areas in eastern China with rich wind energy resources, and are suitable for large-scale wind power development due to the favorable conditions of site construction.

It is not appropriate to build large wind power bases in other regions in China due to the relatively poor conditions of wind energy resources there. Distributed wind power development may be carried out in the light of local resources and construction conditions.

As is seen from wind energy resources and construction conditions, China has a great potential for wind power development, and its installed wind power capacity is likely to surpass 1 billion kW on a long-term basis. An integrated plan shall be made on wind power development and utilization and the power system, on the basis of economic and social development and energy and power demand, so as to achieve scientific and orderly development of wind power, and large-scale development and efficient utilization of wind power through development planning on various power sources and grids. By considering the energy and power development, China's wind power development potential is expected to reach 200–300 GW by 2020 and 400–600 GW by 2030.

4.1.3.5 Development planning and relevant policies

According to China's wind energy resources and construction conditions, in the future, China will give priority to the development of onshore wind power, with offshore wind power as a complement; for onshore wind power, China will give priority to centralized development, with distributed development as a complement. According to the *12th Five-year Plan for Renewable Energy Development* released by the National Energy Administration in August 2012, China will promote large-scale development of wind power in an orderly way in the "Three Norths" area and eastern coastal areas mainly by building large-scale wind energy bases. Meanwhile, according to the *Strategic Action Plan on Energy Development (2014–2020)*, the Chinese government will give priority to the planning and construction of nine large modern wind power bases in Jiuquan, western Nei Mongol, eastern Nei Mongol, northern Hebei, Jilin, Heilongjiang, Shandong, Hami, and Jiangsu as well as supporting power transmission projects. In addition, it plans to energetically develop distributed wind power and take steady steps to develop offshore wind power in southern China as well as the central and eastern regions of the country. By 2020, China's installed wind power capacity is expected to reach 200 GW.

In order to strengthen the management of the grid-connected operation of wind power, China's National Energy Administration released 18 important standards including the *Technical Regulations on Grid-connection Design of Large Wind Power Farm*, on August 5, 2011, involving technical standards that are urgently needed for the development of the wind power industry, such as grid connection of large-scale wind power farms, offshore wind power construction, status monitoring of wind power units, power quality of wind power farms, and requirements on the manufacturing of key wind power equipment. The 18 new technical standards, related to wind power, will change the backward situation of Chinese wind power standards, better guide the development of China's wind power industry, and promote the efficient utilization of wind energy resources. In the future, China will further improve relevant technical standard systems for grid-connected wind power operations, regulate the management of grid-connection technologies, and strengthen the grid-connection test work of wind power units and wind power farms. In December 2011, China released the *Technical Regulations on the Connection of Wind Power Farms to the Power System*, a new national standard for the grid connection of wind power, specifying that all fans used in wind power farms must meet the requirements on grid connection in terms of low-voltage ride through capability, active power control, reactive power control and prediction of wind power, etc. The release and implementation of the standard will effectively enhance the grid-connected wind power quality, and ensure safe and steady operation of the power system and positive development of the wind power industry.

In order to reasonably guide the investments in wind power, ensure healthy and orderly development of the wind power industry, and promote the balanced development of new energy, China's National Development and Reform Commission has issued the *Notice on Improving the Policy on the Benchmark On-grid Price of Onshore Wind Power and PV Power Generation* (F.G.J.G, 2015, 3044), stipulating that the benchmark on-grid price of onshore wind power declines gradually with development scale. In order to make clear the investment expectation, the benchmark prices of onshore wind power for 2016 and 2018 were specified at the same time. The benchmark on-grid price of onshore wind power of in the four types of wind energy resource areas were RMB 0.47, RMB 0.50, RMB 0.54, and RMB 0.60/(kWh) in 2016, and were RMB 0.44, RMB 0.47, RMB 0.51, and RMB 0.58/(kWh) in 2018, respectively. For the onshore wind power projects approved after January 2016 and January 2018, the benchmark on-grid price for the 2 years shall apply. Meanwhile, it is encouraged to determine an onshore wind power project owner and an on-grid electricity price through market competition including invitation for bid, but the on-grid electricity formed as a result of the market competitions must not be higher than the state-specified local benchmark on-grid electricity price for similar projects.

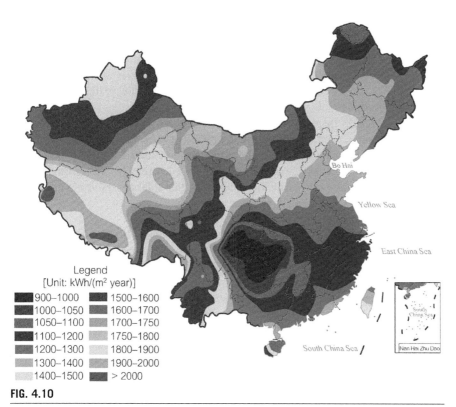

FIG. 4.10

The distribution of solar energy resources in China.

From: Wind and Solar Energy Resources Evaluation Center, China Meteorological Administration.

4.1.4 SOLAR ENERGY

4.1.4.1 Resources and their distribution

China is endowed with abundant solar energy resources. It is estimated[4] that China's land surface receives solar radiation energy of about 1.47×10^7 TWh every year, equivalent to 4.9 trillion tons of standard coal, which is approximately equal to the total electricity generated by more than 10,000 Three Gorges projects. China's solar radiation ranges from 933 kWh/(m²/year) to 2330 kWh/(m²/year), with a median value of 1620 kWh/(m²/year). The distribution of China's solar energy resources is shown in Fig. 4.10.

As is seen from the distribution of China's annual total solar radiation, the solar radiation in Xizang, Qinghai, the central and southern regions of Xinjiang, the central

---

[4]The Institute of Electrical Engineering of the Chinese Academy of Sciences, *Research on the Prospect of Development and Utilization of China's Solar Energy Resources*, November 2009.

and western parts of Nei Mongol, Gansu, Ningxia, western Sichuan, Shanxi, and northern Shaanxi is very large.

The main characteristics of the distribution of China's solar energy resources: both the high-value center and the low-value center of solar energy are located between 22°N and 35°N. The Qinghai-Xizang Plateau is a high-value center, while the Sichuan Basin is a low-value center; the annual amount of solar radiation in western China is higher than that in eastern China, moreover, except for Xizang and Xinjiang, the annual amount of solar radiation in southern China is basically lower than that in northern China; most southern regions of China are located between 30° N and 40°N, and are cloudy, foggy, and rainy. In such regions, the distribution of solar energy does not comply with the general rule that solar energy varies positively with latitude, that is, the solar energy rises as the latitude increases, rather than dropping as the latitude increases.

In accordance with the solar radiation it receives, China is divided into five major regions, as is shown in Table 4.11. Specifically, the annual sunshine hours in Categories 1, 2, and 3 regions are greater than 2200 h, and the annual amount of solar radiation there is higher than $1390 \, kWh/(m^2/year)$. The regions have abundant or relatively abundant solar energy resources and cover a large area that accounts for approximately 2/3 of the total area of the country. Specifically, the annual sunshine hours of the Qinghai-Xizang Plateau in northwest China, northern Gansu, northern Ningxia, and southern Xinjiang are between 3200 and 3300 h, with an annual amount

Table 4.11 Category of China's solar radiation (by province)

Category of region	Sunshine duration (h)	Annual solar radiation [kWh/(m²/year)]	Provinces (regions)
Category 1	3200–3300	1860–2330	Qinghai-Xizang Plateau, northern Gansu, northern Ningxia, southern Xinjiang
Category 2	3000–3200	1630–**1860**	Western Nei Mongol, southern Ningxia, central Gansu, eastern Qinghai, southeast Xizang, northern Xinjiang, and the northwest part of Hebei
Category 3	2200–3000	1390–1630	Shandong, Henan, southeast Hebei, southern Shanxi, Jilin, Liaoning, Yunnan, northern Shaanxi, southeast Gansu, southern Guangdong, southern Fujian, northern Jiangsu, northern Anhui
Category 4	1400–2200	1160–1390	Fujian, Zhejiang, and some areas of Guangdong
Category 5	1000–1400	933–1160	Central and eastern parts of Sichuan, Guizhou

of solar radiation of 1860–2330 kWh/(m^2/year), equivalent to the heat from burning of 225–285 kg standard coal. These regions have the most abundant solar energy resources of China and have good conditions for the utilization of solar energy.

4.1.4.2 Development situation

In recent years, driven by the international solar energy market, China's photovoltaic (PV) cell manufacturing industry has seen rapid development. Its solar energy PV cell manufacturing industry has formed a relatively complete industrial chain ranging from crystalline silicon material and battery production to modular packaging and manufacturing of special production equipment. The output of solar PV cell keeps growing rapidly, the PV power generation market has taken shape, the solar thermal power generation market has emerged, and the solar-thermal utilization market keeps expanding.

Installed solar power capacity. In recent years, China has sped up its intensive development of solar energy, started the construction of 10–20 MW grid-connected solar PV power generation bases in Gansu and Qinghai and the pilot 10 MW solar thermal power generation demonstration project. As at the end of 2015, the installed grid-connected solar power capacity of China had reached 41.58 kW, 160 times as at the end of 2010, with an average annual growth rate of 176% during the "12th Five-year Plan" period, indicating that solar energy has moved from the initial stage to the large-scale development stage.

Solar thermal power generation technology. In recent years, due to the development situation of and polices on renewable energy at home and abroad, China's solar thermal power generation market has emerged, and many enterprises are conducting preliminary research. Nevertheless, in terms of efficiency and economy of solar thermal power generation technology, China still lags far behind developed countries. China is a latecomer to research on solar thermal power generation technology, but has made great headway in slot- and tower-type solar thermal power generation technologies thanks to its previous technological accumulation in the fields of material, key devices, and substation design. The work related to invitations for bidding for the franchise of the 50 MW slot-type solar thermal power generation station in Ordos, China's first commercial solar thermal power generation project, was completed in April 2011. The 1.5 MW Badaling tower-type photothermal demonstration project of the Institute of Electrical Engineering of the Chinese Academy of Sciences, China's first MW-level solar thermal power generation project, was put into operation officially in November 2012. In addition, in Ningxia and Xizang, there are about 185 MW slot-type thermal power generation projects at the stage of preliminary feasibility study or at the stage of feasibility study.

4.1.4.3 Development potential

From the perspective of resource, China has great potential in the development of solar PV power generation in the future. China has numerous building roofs. In northwest China where there are rich solar energy resources, there are large-area deserts and wastelands that can be used for solar energy development. It is roughly estimated that China has now a total of about 40 billion km^2 of roofs. Assuming that 1% of the roofs are installed with a PV system, the total installed PV power capacity can reach about 35.5–66.2 GW, with an annual electricity output of 2.87–5.43 TWh.[4] China has about 2.64 million km^2 of deserted lands, of which the deserted lands in arid regions cover more than 2.5 million km^2 and are mainly distributed in northwest China with rich solar resources. Assuming that 3% of the Gobi and other deserts are utilized in China, the installed capacity of potential available solar energy may reach 2.7 billion kW, with an annual power generation of 4.1 trillion kWh.

4.1.4.4 Development planning and relevant policies

According to the *Strategic Action Plan on Energy Development (2014–2020)*, the Chinese government will boost the construction of PV bases in an orderly way, build channels for local consumption and utilization and centralized power transmission, speed up the building of demonstration zones for the application of distributed PV power generation, and take steady steps to implement solar thermal power generation demonstration projects. Also, it will offer better grid-connected solar power generation services, encourage distributed PV power generation by using the rooftops of large-scale public buildings, facilities, and industrial parks. By 2020, the installed PV capacity of China is expected to reach about 100 GW.

In recent years, the Chinese government has supported PV power generation, grid connection, and consumption in a sustained manner. However, under the circumstances of rapid expansion of the PV power generation market, unjust resources allocation, irregular management, speculations, and profiteering began to project investment and development.

In June 2014, China's National Energy Administration issued the *Notice on Strengthening the Construction and Operation Management of PV Power Stations* (exposure draft), in which it makes the development and planning of large PV power station bases an important part of national power station development planning annually and for the future. Considering the current situation of difficulties in power transmission from ground-based solar PV power stations, it proposes that power grid companies operate in coordination with power transmission projects.

In September 2014, the National Energy Administration issued the *Notice on Further Implementing the Policies on Distributed PV Power Generation*, for the purpose of using building roofs and auxiliary sites to carry out distributed PV power generation projects. When a project is put on record, either the mode of "generating electricity for self-use and connecting surplus electricity to the grid" or "connecting all generated electricity to the grid" may be adopted. If a solar PV power station is built on the ground or in an agricultural greenhouse where there are no facilities for electric power consumption, connected to the grid at a voltage level of 35 kV or below (66 kV and below for northeast China), has a capacity of up to 20 MW and the electricity generated is mainly used for consumption at grid-connected power transformation areas, it shall be included in the management system of distributed PV power generation, and be subjected to the benchmark on-grid price for local

PV power stations. Moreover, prefecture- or county-level grid companies shall handle the matter of grid connection and provide corresponding grid-connection services in accordance with the simplified procedures, and on the basis of the stipulations of Article 17 of the *Interim Procedures for the Management of Distributed Power Generation* and established "green channels."

In December 2015, China's National Development and Reform Commission issued the *Notice on Improving the Policy on the Benchmark On-grid Price of Onshore Wind Power and PV Power Generation* (F.G.J.G, 2015, 3044), to adjust the policies on the benchmark on-grid electricity price of new PV power stations. In order to make clear the investment expectancy, the document specifies that the benchmark price of PV power in 2016 and the price after 2017 shall be set separately. In 2016, the benchmark on-grid electricity price for PV power stations at the three types of PV resource areas was RMB 0.80, RMB 0.88, and RMB 0.98/(kWh), respectively. The PV power generation projects that were put on record and included into the annual scale management system after January 1, 2016 is subject to the benchmark on-grid electricity price of PV power generation in 2016. Meanwhile, it specifies that if a distributed PV power generation project is based on building roof and auxiliary sites, when it is put on record, either the mode of "generating electricity for self-use and connecting surplus electricity to the grid" or "connecting all generated electricity to grid" may be adopted. For the projects that are carried out in the mode of "generating electricity for self-use and connecting surplus electricity to the grid," in circumstances of significant reduction of electric load (including disappearance), or the failure in implementing the relationship between the power supply and consumption, it is permitted to be changed into the mode of "connecting all generated electricity to the grid." The electricity of the projects in the mode of "connecting all generated electricity to grid" shall be purchased by power grid enterprises as per the benchmark on-grid electricity price of local PV power stations. If the mode of "connecting all electricity to grid" is adopted, the project owner shall make an application to local competent energy authorities for record change, and then is not allowed to return to the mode of "generating electricity for self-use and connecting surplus electricity to the grid."

In the future, the Chinese government will further promote market competition mechanisms for large-scale solar energy development, establish a mode of solar-based power operation, strengthen the planning and project management of solar energy, improve the technical regulations and standard systems for solar energy, and strengthen the industrial management of the PV manufacturing industry.

4.1.5 **BIOMASS ENERGY**

4.1.5.1 Resources and their distribution

(1) Resources

China's biomass energy resources are mainly crop stubble, tree branches, excrement from livestock, energy-producing crops (plants), industrial organic wastewater, urban sewage and wastes, most of which are crop stubble, tree branches and forestry waste, energy-producing crops, and urban waste. No systematic evaluation has been made on biomass energy resources. Currently, the annual output of China's biomass raw material resources is equivalent to about 460 million tons of standard coal, of which the annual output of crop stubble and wood residue is equivalent to about 170 and 200 million tons of standard coal, accounting for 37.0% and 43.5%, respectively, see Table 4.12 for details.

(2) Distribution

China's biomass energy resources are mainly distributed in the eastern coastal areas, the mountainous area and rural areas in southwest China and the "Three Norths" areas. China's crop stubble resources are mainly distributed in Henan, Shandong, Hebei, Anhui, Jiangsu, Heilongjiang, Sichuan, and Jilin; the tree branch resources are mainly in northeast and southwest regions. The domestic/industrial waste resources are mainly in industrially developed regions including Shanghai, Jiangsu, and Zhejiang. Bagasse from tuber crops for power generation is mainly in southwest China.

4.1.5.2 Development situation

The main methods of biomass energy utilization are biomass power generation, biomass gasification (including biogas), biomass briquette fuel, and biological liquid fuel. During the "12th Five-year Plan" period, China's biomass energy developed rapidly. Biomass was used for power generation as well as making liquid fuel, fuel gas, and briquette fuel, the utilization scale was expanded, the technologies kept advancing and preliminarily industrialized in some areas, playing a positive role in the replacement of fossil energy, promoting environmental protection and increasing farmers' income, etc.

Biomass power generation. At the end of 2014, China's installed capacity of grid-connected biomass power generation stood at 9.48 GW, including that from agriculture and forestry biomass power generation at 5 GW, that from garbage power generation at 4.24 GW, that from methane power generation at 220 MW, and that from gas power generation at 20 MW. The biomass power generation has formed a certain scale, with an annual power generation of over 40 TWh. China has made rapid progress in biomass power generation technology and equipment manufacturing, and has mastered high-temperature and high-pressure biomass power generation technologies.

Biological liquid fuel. In 2014, China's biological liquid fuel output was approximately 2.1 million tons. A batch of new varieties of energy crops with strong stress tolerance and high yield have been cultivated, the technologies for production of cassava and sweet sorghum ethanol have been basically matured, great progresses have been made in research and development of cellulosic ethanol technology, and several small-scale test units have been built. Meanwhile, new progress in China's independently developed aviation biofuel has been made in test flights, and there has been a good development momentum of R & D innovations in biological liquid fuel.

Table 4.12 China's biomass raw material resources (unit: 10,000 ton)

Source	Available resources		Utilized resources		Remaining available resources	
	Physical quantity	Standard coal equivalent	Physical quantity	Standard coal equivalent	Physical quantity	Standard coal equivalent
Crop stubble	34,000	17,000	800	400	33,200	16,600
Processing residues of agricultural products	6000	3000	200	100	5800	2900
Wood residues	35,000	20,000	300	170	34,700	19,830
Excrements fromof livestock	84,000	2800	30,000	1000	54,000	1800
Urban wastes	7500	1200	2800	500	4700	700
Organic wastewater	435,000	1660	2770	10	432,300	1590
Organic waste residue	95,000	440	4800	20	90,200	380
Total		46,000		2200		43,800

Note: The above data come from the 12th Five-year Plan on Biomass Energy Development from the National Energy Administration, July 2012.

Biomass gas. In 2013, there were about 41.22 million rural households using biogas in China, with an annual output of biogas of about 13.7 billion m^3. More than 50,000 livestock and poultry farms, designed for biogas generation, have been built, with an annual output of about 800 million m^3 of biogas. In rural areas, the biogas technology continues to mature, the industrial system improves gradually, and the system for property management biogas services has been established in many places. The technology and process for biomass gasification and centralized gas supply have kept improving, and about 1000 biomass projects for centralized gas supply have now been built and operated.

Biomass briquette fuel. In 2014, about 7 million tons of biomass briquette fuels were produced and they were mainly used for rural residents and urban heating boilers, as well as providing raw materials for making biomass charcoal. The briquette fuel equipment features a significant reduction of energy consumption, obvious life extension for parts wearing out, and a marked improvement of maintainability. Furthermore, China has the preliminary conditions for large-scale industrialized development of briquette fuel.

In recent years, China's biomass direct combustion power generation technology had become relatively mature and has become an effective technology for the consumption and utilization of crop stubble in regions with these resources. On the basis of introducing foreign waste incineration power generation technologies and equipment, China basically has the ability for manufacturing such equipment, and it has used imported foreign equipment and technologies to carry out some demonstration projects for power generation by landfill gas. However, on the whole, China still lags behind countries at an international advanced level in terms of raw material collection, purification treatment, and combustion equipment manufacturing for biomass power generation.

4.1.5.3 Development potential

Compared with currently prevailing modes of power generation such as coal-fired power and hydropower, etc., biomass power generation has no cost advantages, and compared with such future technologies of electricity generation from such renewable energy sources as wind energy and solar energy, biomass power generation has no resource advantages. China is poor in oil and natural gas resources. Therefore, from the perspective of an approach to resource utilization, if any breakthrough has been made in the second-generation biomass fuel ethanol technologies, in the future, the production of biomass fuel ethanol shall be emphasized for energy conversion and utilization of biomass resources, so as to partially replace transportation fuel. Considering the limited availability of biomass resources, economical efficiency of products, and the competition between the use of biomass resources and other resources, biomass power generation shall be restricted and only conditionally developed. China has various raw materials for biomass power generation, which have distinct characteristics, include mostly agricultural waste, as well as wood residue and potential shrub wood. Therefore, different technical plans shall be made in line with local conditions so as to develop biomass power generation

Table 4.13 The maximum development scales, for which China's biomass resources can support

Area of application		Maximum development that biomass resources can support		Annual energy output	
		Quantity	Unit	Quantity	Unit (GWh)
1.	Biomass power generation	10,021	10,000 kW	6013	100
	Agriculture and forestry biomass power generation	8345	10,000 kW	5007	100
	Biogas power generation	1028	10,000 kW	617	100
	Garbage power generation	649	10,000 kW	389	100
2.	Gas supply based on biomass	2310	100 million m^3		
	Methane	1982	100 million m^3		
	Gas based on agricultural wastes	128	100 million m^3		
	Methane from industrial wastewater	200	100 million m^3		
3.	Biological briquette fuel	10,431	10,000 tons		
4.	Biological liquid fuel	5215	10,000 tons		
	Biofuel ethanol	4172	10,000 tons		
	Biodiesel and aviation fuel	1043	10,000 tons		

Note: *Measured on the basis of relevant data of the* 12th Five-year Plan on Biomass Energy Development.

technologies and encourage diversified development of those technologies. Medium and small-scale biomass power generation shall be given priority, on the basis of the supply of raw materials. In accordance with China's potential for biomass energy resources, the estimate of the maximum development scale of power generation, fuel, and gas that China's biomass energy resources can support, is shown in Table 4.13.

4.1.5.4 Development planning

In December 2014, China's National Development and Reform Commission issued the *Notice on Strengthening and Regulating Relevant Requirements on the Biomass Power Generation Project Management* (F.G.B.N.Y, 2014, 3003), laying the stress on encouraging the development of combined heat and power generation

Table 4.14 China's installed capacity of for biomass power generation (unit: 100 GW)

Region	2014	2020
North China	248	700
East China	297	350
Central China	185	570
Northeast China	106	330
Northwest China	10	170
Southern China	102	880
Total	948	3000

(cogeneration) based on biomass, requiring that cogeneration is adopted for new biomass power generation projects, and that cogeneration renovations are made for existing biomass power generation projects, in accordance with the heat market and conditions of technical and economic feasibility. The notice has further defined the level and scope of approval of biomass power generation projects, regulated management procedures, showed the direction of the development of the biomass power generation industry, and provided the industrial management and sustainable development of biomass power generation with strong policy supports.

During the "13th Five-year Plan" period, the Chinese government will speed up the large-scale development and utilization of biomass energy, on the basis of the local conditions of biomass resources and characteristics of local energy demands, work faster to promote biomass energy utilization of biomass power, biomass gas, briquette fuel, and liquid fuel, for which the application technologies are basically mature, and the conditions and basis for industrialization have been created, and boost the large-scale and industrial development of biomass energy. From the aspects of power generation and application of biomass energy, the Chinese government aims to achieve growth in a reasonable and orderly way, particularly from the aspects of agriculture and forestry biomass power generation, garbage power generation and biomass gas power generation, etc. Table 4.14 shows the distribution of China's biomass power stations in accordance with the conditions and collection of energy resources for biomass power generation.

4.2 **FOSSIL ENERGY**

4.2.1 **COAL**

4.2.1.1 Capability of sustainable supply of coal

China has relatively abundant coal resources. It is predicted that China has 5.6 trillion tons of coal resources with a vertical depth of less than 2000 m and an available reserve of about 1.1 trillion tons. Compared with major coal-producing countries in the world, China has relatively poor conditions in the amount of coal resources, and few coal resources suitable for open-cast mining. Its coal-rich western and northern regions have better conditions for mechanical mining, but the development of coal resources there is restricted due to the fragile ecological environment and insufficient water resources.

China's coal is mainly produced in Shanxi, Shaanxi, and western Nei Mongol, but is mainly consumed in the coastal areas in southeast China. The pressure caused by "coal transportation from the north to the south and from the west to the east" keeps increasing, and long-distance transportation has become a bottleneck in the supply and demand of China's coal. First, inadequate transportation capabilities, due to the relatively low growth of the railway transportation, the poor situation of coal transportation has not been eased, intensifying the shortage of coal supply. Second, coal transportation is not cost effective. As resources in North China and East China are on the verge of exhaustion, it is imperative to develop coal resources in Xinjiang and some other provinces in western China at an appropriate time. However, Xinjiang is far away from central and eastern China, and the lignite resources in Xilinguole League are not suitable for long-distance transportation. As is seen from the development tendency, coal shall be consumed locally by means of the construction of pithead power plants so as to resolve the problem of long-distance transportation.

In addition, China's coal mines are plagued with potential safety hazards and poor working environments, which hinder the sustainable development of the coal industry. At present, China's coal mining accidents directly cause the deaths of over 2000 persons a year, several times that of developed countries, in terms of death rate per ton of coal. As the concept of "putting people first" takes root among the people, safe coal production has become an important precondition for coal production capacity, and workplace safety has become a constraint condition of China's coal production capacity.

Under the circumstances of giving overall consideration to coal resources, geological conditions, ecological environments, water resources, transportation economy, and natural hazards, the maximum scale of coal development in China will be about 4.1 billion tons/year in 2030 and 4 billion tons/year[5] in 2050, as is shown in Fig. 4.11. According to the requirements on non-fossil energy use accounting for 15% of China's total energy consumption, by 2020, the energy supply for non-fossil energy is expected to reach about 70–80 million tons of standard coal. It may help to control China's total coal supply between 4 and 4.3 billion tons by 2020. In 2020, the proportion of coal to primary energy consumption will drop from 65% to less than 60%, and the proportion of coal supply to energy supply will decline gradually, thereby lessening the reliance on coal resources, safeguarding China's energy supply security, and reducing the pressure on the ecological environment.

---

[5]Scientific Development 2030—Research Report on China National Energy Strategy, National Energy Administration, 2010.

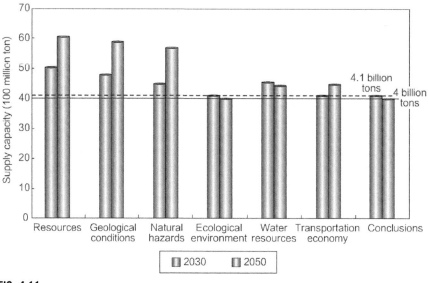

FIG. 4.11

China's capability of coal supply under the six constraints.

From: Scientific Development 2030—Research Report on China National Energy Strategy,
National Energy Administration, 2010.

4.2.1.2 Development potential of coal-fired power base

For a long time, China has been transporting coal to its central and eastern regions for the construction of coal-fired power stations for power generation. The electric power transmitted there yields a very small amount, which is one of the major reasons for China's long-standing inadequate capability problem regarding coal transportation and its serious environmental pollution in the central and eastern regions. The research results show that, building large-scale coal-fired power bases in the coal producing areas in the western and northern regions, delivering coal-fired power to eastern and central regions via ultrahigh voltage (UHV) grid, and building a comprehensive energy transport system for "coal transportation and power transmission" plays an irreplaceable role in improving China's energy structure, reducing the power supply cost of the central and eastern regions, increasing the level of energy supply security, optimizing the utilization of China's environmental resources, raising the efficiency of the comprehensive utilization of energy resources, promoting the coordinated development of regional economy and saving the valuable land resources, etc.

China is endowed with very abundant coal resources in its western and northern regions, and its coal-producing areas including Shanxi, Northern Shaanxi, Binchang, Ningdong, Western Nei Mongol, Xilinguole League, Hulunbeier League, Hami, and Zhundong have resource conditions for building large-scale coal-fired power bases. Within the bases, except for Hulunbeier League, these coal producing areas are lacking in water resources. In accordance with the relevant results of the *In-depth*

Table 4.15 Coal power base power source development potential

Name of base	Available reserves of coal (100 million tons)	Water resources (100 million m^3)	Installed capacity of recoverable power source (10,000 kW)
Shanxi	2663	123.8	10,000
Northern Shaanxi	1291	48.4	4380
Ningdong	309	3.2	4880
Jungar	256	3.6	6600
Ordos	560	25.8	6600
Xilingol League	484	26.1	5000
Hulunbeier League	338	127.4	3700
Huolinhe	118	2.4	1420
Hami	373	5.7	2500
Zhundong	789	13.9	3500
Yili	129	170	8770
Binchang	88	15.1	1440
Longdong	142	12.5	2660
Huainan	139	58	2500
Total	7679	635.9	62,640

Research on Energy Base Construction and Medium and Long-term Development Planning of the Electric Power Industry, by using water-saving and air-cooled units, on the basis of enhancing the construction of water conservancy projects and strengthening reutilization of urban reclaimed water and mine drainage, predictions have been made on the basis of water resources supply and demand for different provinces, regions, and industries, that by 2020, the water resources for power generation in coal producing areas will be capable of supporting air-cooling coal-fired power generation units, with a total capacity of 600 GW, and can thus meet the needs of the construction of large coal-fired power bases and large-scale coal-fired power transmission. The power source development potentials of the coal-fired power bases are shown in Table 4.15. Actually, only about 5% of the total water supply is allocated for power generation, and the remaining water supply can meet the water demands of other industries of the regions as well as for economic and social development.

4.2.1.3 Development planning

According to the *Strategic Action Plan on Energy Development (2014–2020)*, on the principles of "Safety, Green, Intensiveness, and Efficiency," the Chinese government plans to speed up the development of technologies for the clean development and utilization of coal and to keep raising the level of this development and utilization.

Clean and efficient development of coal-fired power. The Chinese government will change the method of coal consumption, work hard to increase the proportion of intensive and efficient coal-fired power generation, and improve the criteria for approval of coal-fired power units, so as to ensure that the coal consumption of new coal-fired power generation units is lower than 300 g standard coal per kWh and the pollutant emission approaches the established emission level of generated gas.

Boost the construction of large bases and large corridors for coal-fired power development. According to the characteristics of regional distribution of water resources and the ecological and environmental bearing capacity, the Chinese government plans to adopt strict standards for environmental protection and safety of coal mines, promote such green mining technologies as backfilling and water retention, give priority to the construction of 14,100 million ton coal bases including Jinbei, Jinzhong, Jindong, Shendong, northern Shaanxi, Huanglong, Ningdong, Luxi, Lianghuai, Yungui, Jizhong, Henan, eastern Nei Mongol, and Xinjiang coal bases. By 2020, the output of the above-mentioned bases is expected to account for 95% of the total output of China. The most advanced energy-saving, water-saving, and environment-friendly protection power generation technology will be used, the nine key 10 million-kW coal-fired power bases including Xilinguole, Ordos, Jinbei, Jinzhong, Jindong, Northern Shaanxi, Hami, Zhundong, and Ningdong will be built. Long-distance and large-capacity power transmission technologies will be developed, the scale of West-East Power Transmission will be expanded, and the North-South Power Transmission Project will be implemented.

Also, the Chinese government will strengthen the construction of railways for coal transportation, give priority to the construction of railways between western Nei Mongol and central China for coal transportation, and improve the corridors for transporting coal from western China to eastern China. By 2020, China's railway transportation capacity of coal is expected to reach 3 billion tons.

The Chinese government will raise the level of clean coal utilization, formulate and implement a plan on clean and efficient utilization of coal, vigorously advance cascading utilization of coal with different grades and quality, raise the proportion of coal washing and screening, encourage in situ conversion and utilization of low caloric value coal and low-quality coal gangue. The Chinese government will establish a sound system for coal quality management, strengthen the supervisory management of coal development, processing, transformation, and use, and strengthen supervision of the quality of coal, significantly reduce distributed and direct coal combustion, and encourage the use of clean coal and briquette coal in rural areas.

4.2.2 OIL

China is a country that is deficient in oil and natural gas. It lacks conventional oil resources and it is hard to achieve significant and sustained growth from its output of crude oil. According to the new round of oil and gas resources evaluation, as of the end of 2008, China has only 2.1 tons of recoverable reserves of oil, which accounts for merely 1.2% of that in the world, and its oil resources are seriously inadequate.

Among the remaining recoverable reserves of oil, low-permeation or ultra-low permeation oil, heavy oil, thickened oil, and oil with a burial depth more than 3500 m account for over 50% of the total, while most recoverable resources that have not yet been proven are hard to recover due to greater burial depth, lower quality, and stronger marginality. The occurrence characteristics of oil resources determine that it is hard for China to increase its oil resource reserves and that the exploration cost will further rise. By giving comprehensive consideration to such factors as resources, development technologies, and economic risk, for the purpose of ensuring long-term stability of oil output, the Chinese government will appropriately control the mining speed and keep the peak output of crude oil to under 200 million tons. China has a low oil reserves to production (RP) ratio, and in the future, as the proportion of energy consumption for transportation increases, the proportion of oil consumption will gradually go up, and its foreign trade dependency will also increase. Therefore, we shall work hard to save resources, strengthen exploration, take active steps to seek for alternative energy sources, and import an appropriate amount of energy, so as to ensure that the external dependency on oil is within a reasonable limit.

(1) China's conventional oil resources are limited and it is hard to realize continuous and significant growth of the output of crude oil.

It is predicted that the peak value of newly added oil reserves of China will be 180–200 million tons, and long-term and steady supply can be ensured on such a scale. By 2050, the growth of China's recoverable oil reserves is expected to reach 110–130 million tons. The new proven recoverable oil reserves of China between 2006 and 2050 is expected to reach 7.7 billion tons, of which the total new proven recoverable reserves of the six major basins including Songliao, Bohai Gulf, Jungar, Tarim, Ordos, and Bohai Gulf is expected to reach 5.9 billion tons, accounting for 76% of the country's total.

As is seen from different development plans, in the case of production based on the rapid development plan, namely that all reserves in the new areas are utilized, and assuming that the RP ratio remain at 9.3, China's oil output is likely to reach its peak around 2020, that is, an annual output of 230 million tons, but will drop rapidly, even to about 160 million tons by 2050, as is shown in Fig. 4.12. If China controls its mining speed appropriately and partially uses the reserves in the new areas, the RP ratio will rise overtime and reach 10.3 by 2020, then remain stable and achieve balanced development. If China controls its peak output of crude oil under 200 million tons, it may still reach 198 million tons by 2030 and will drop slightly to 171 million tons by 2050. From the perspective of sustainable development of China's oil industry, if China adopts a plan on reserves and keeps the peak output of oil to around 190 million tons, it may extend the peak output beyond 2030 and even maintain an output level of 180 million tons by 2050, which is more favorable for safeguarding China's oil security and national economic development.

By region, in terms of the petroleum production by 2030, eastern China will continue to take the lead, the status of western China will be on the rise, and the

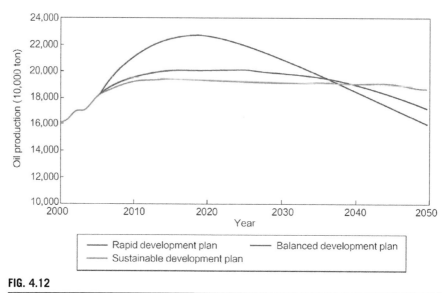

FIG. 4.12

The predictions on China's oil output under three plans.

From: Scientific Development 2030—Research Report on China National Energy Strategy,
National Energy Administration, 2010.

oil output of western China is expected to exceed that of eastern China after 2030.

As is seen from the supply of oil, the output from new proven reserves will play a leading role in oil output in the future and is expected to account for over 75% of the total output of oil around 2030. Therefore, the key to achieving China's goal of oil output in the future is to strengthen oil exploration, find more proven reserves and ensure continuous investment in, and exploration of, proven reserves.

(2) Nonconventional oil resources have low grade and play a complementary role in conventional oil supply.

According to the new round of evaluation of oil and gas resources in China,[6] China has a total of 243.2 billion tons of oil shale recoverable resources, equivalent to 11.98 billion tons of recoverable shale oil and 2.26 billion tons of sand oil. The detailed distribution of the resources is shown in Table 4.16. However, China's shale oil and sand oil resources are of a low grade. According to the domestic and overseas research, the content of shale oil that can be commercially recovered is generally larger than 8%–10%, and oil shale with an oil content of less than 10% is generally used for power generation and

---

[6]The oil and gas resource evaluation jointly conducted by the Ministry of Land and Resources of China and the National Development and Reform Commission of China in 2003.

Table 4.16 Grades of oil shale and oil sand resources of in China

Oil shale			Oil sand		
Grade (oil content, %)	Shale oil (100 million tons)	Proportion to China's total (%)	Grade (oil content, %)	Shale oil (100 million tons)	Proportion to China's total (%)
3.5–5	38.5	32	3–9	11.7	52
5–10	46.7	39	6–10	10.5	46
>10	34.6	29	>10	0.4	2
Total	119.8	100	Total	22.6	100

comprehensive utilization of cement; the oil content of sand oil that can be commercially recovered is generally greater than 10%.

China's oil shale resources are widely distributed in 47 basins of its 20 provinces, mainly in Songliao, Ordos, Lunpola, Jungar, Qiangtang, Maoming, and Qaidam, which have a total of 11.29 billion tons of recoverable shale oil, accounting for 94% of the country's total.

At present, the shale oven technology has become increasingly matured and more and more large-scale shale ovens are being adopted in the world. It is predicted that the technology will be commercialized around 2015. The predicted change trend of the output of China's nonconventional oil is shown in Table 4.17. According to the conditions of China's oil shale resources and the technological development tendency in the future, shale oil output may reach 3 million tons by 2020 and 6 million tons by 2030.

At present, the world's mining technologies for the opencast mining of oil sand resources are relatively mature, and according to China's conditions of oil sand resources and the technical development tendency in the future, it is predicted that the sand oil output will reach 1.5 million tons by 2020 and 2.5 million tons by 2030.

On the whole, China's supply capability of nonconventional oil is limited. It is estimated that its total output of shale oil and sand oil will be merely 4.5 million tons in 2020 and 8.5 million tons in 2030, playing a complementary role in the entire system of oil supply.

Table 4.17 Predictions of the growth tendency of China's nonconventional oil output

Category	2020	2030
Shale oil	300	660
Sand oil	150	250
Total	450	850

(3) Since its oil resources mainly play a role of basic support, China will keep increasing its import of oil.

Before 2030, China will be built on its conventional oil resources, focus on the eight proven onshore and offshore petroliferous basins, strengthen oil and gas exploration and the development of detailed exploration in lithological, foreland, marine, and mature basins, and ensure the steady growth of its oil reserves and crude oil output. The output of China's conventional oil will remain at about 200 million tons, and the output of nonconventional oil including that of shale oil and sand oil will reach about 8.5 million tons.

Moreover, China will deploy its oil and gas resources around the world in a diversified way, further promote international cooperation in the oil and gas field, raise its capability of utilizing foreign oil and gas resources, develop a diversified and stable system for oil trade, and ensure the steady oil and gas supply and safety of corridors for oil delivery. The total import and export of oil is expected to exceed 500 million tons by 2030, and its external dependency on oil will further increase from 60% in 2015 to 73% in 2030.

4.2.3 NATURAL GAS

Natural gas is a kind of high-quality, efficient, clean, and low-carbon fossil energy. As a result of energy conservation, emission reduction, deepening of energy structural adjustment, and increasing urbanization, the proportion of natural gas to China's primary energy consumption structure will keep increasing. In accordance with the *Strategic Action Plan on Energy Development (2014–2020)*, China will accelerate conventional natural gas exploration and development, and strive to build eight large natural gas production bases with an annual output of more than 10 billion m^3. By 2020, China aims to increase its proven geological reserve of conventional natural gas by 5.5 trillion m^3, and achieve an annual output of conventional natural gas of 185 billion m^3. Meanwhile, it will speed up the construction of natural gas pipelines, and make the length of the main natural gas pipelines exceed 120,000 km by 2020.

China is relatively lacking in conventional natural gas resources, relies heavily on the import of gas, and for this faces great uncertainties in gas supply and price. Therefore, an overall consideration shall be given to such factors as supply, market, and price for the utilization of natural gas resources, priority shall be given to people's livelihood, and natural gas resources shall be mainly used for industrial fuel as well as civil and chemical uses. In the aspect of gas power generation, priority shall be given to distributed gas cogeneration, and large gas-fired power stations shall be developed appropriately.

(1) It is hard for a homemade natural gas supply to meet the rapidly growing consumption needs of China.

According to the results of the latest round of oil and gas resource evaluations, China has 22 trillion m^3 of recoverable natural gas. On the basis of

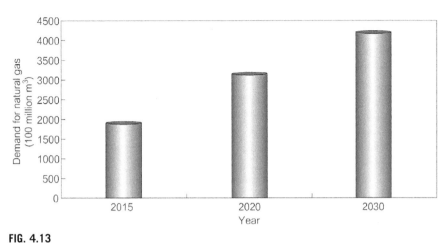

FIG. 4.13

Prediction of China's demands for natural gas between 2015 and 2030.

the statistical relationship between peak output and recoverable resources of natural gas in highly developed countries and regions, it is estimated that China's recovery rate of natural gas is around 1.1% during the period of peak output, with an output of 240 billion m^3 a year. On the basis of the predictions of the output of dissolved gas and coal-bed gas, China's natural gas output is expected to reach 240 and 300 billion m^3 by 2020 and 2030, respectively, while in the same period the demands for natural gas will reach 315 and 420 billion m^3, respectively. Particularly, after 2020, the growth of China's supply capability of natural gas will slow down, while the consumption demand will continue to grow at a high rate, and the supply-demand gap will continue to widen, reaching approximately 120 billion m^3 by 2030.

The predictions of China's demand for and supply of natural gas between 2015 and 2030 are shown in Fig. 4.13 and Table 4.18.

(2) It will be more difficult for China to explore and develop its natural gas in the future.

In the future, China will gradually turn its natural gas exploration and development from land to sea, and raise the complexity and depth of natural gas

Table 4.18 Predictions of China's demand for and supply of natural gas (unit: 100 million m^3).

Year	Demand	Domestic supply	Self-sufficiency ratio (%)
2215	1910	1318	69.0
2220	315	2440	76.2
2030	4200	3000	71.4

exploration. The exploration targets feature complicated surface and underground conditions, early reservoir formation, great burial depth (greater than 3000 m in most cases), poor conditions (low permeability), and complex types (composite type in most cases), giving rise to frequent earthquakes as well as higher difficulty and costs of exploration, including drilling in new areas. In old areas, there has been no important discovery for a long time after the breakthrough made during the early period. The above factors will restrict the process of China's natural gas exploration to a certain extent, thereby affecting the realization of the target of resource exploration.

To greatly increase its natural gas output, China must focus on the highly efficient development of complicated gas deposits featuring super-depth, low permeability, high sulfur content and volcanics, as well as effective development and production of coal-bed gas. However, some technical problems related to the development and supporting engineering processes of such resources still need to be urgently tackled. From the aspect of safe development of sulfur gas fields, China is still lacking practical experience and mature technologies such as the technology for complete well gas extraction in sour gas fields and auxiliary technologies for corrosion prevention in gas wells with high sulfur content. From the aspect of economical and effective development of low-permeability and low-yield gas fields, it is urgently necessary to address a series of engineering and technological issues relating to increasing single-well production capacity, recovery ratio and reserve utilization level, as well as reducing costs, etc. From the aspect of rapid and effective development of coal-bed gas, we need to strengthen optimization of high-yield block and reservoir formation, dynamic analysis on coal-bed gas, research into policies for development technologies, and evaluation of technological effects for production increase, etc. The above technologies are key restriction factors of gas reservoir development. The realization of the natural gas development goal will be severely affected if no breakthrough or progress is made as scheduled.

(3) Disadvantaged geographical location will make it difficult for dispatching of natural gas resources.

China's natural gas resources are mostly distributed in central and western China, and the natural gas resources in Tarim, Ordos, and Sichuan basins account for 52% of the country's total (specifically, Tarim basin accounts for 20%, Ordos basin accounts for 19%, and Sichuan basin accounts for 13%); and the natural gas resources in other basins account for 48% of China's total natural gas resources.

It is predicted that by 2030, the resources in the three major basins (Tarim Basin, Ordos Basin, and Sichuan Basin) will account for 68% of China's total output of natural gas (specifically, the natural gas resources in Tarim Basin accounts for 30% of the total, that in Ordos Basin accounts for 16%, and that in Sichuan Basin accounts for 22%); while the resources in other basins account for 32% of the total resources of China. The major areas of natural gas consumption are the Yangtze River Delta, the Pearl River Delta, and the Circum-Bohai-Sea region in the central or eastern regions of China. The inaccessibility of resources for the target market has put forward higher requirements for natural gas pipelines and the building of the peak shaving capability, and posed severe challenges to flexible dispatching and steady supply of natural gas.

(4) China will gradually increase its importing of natural gas, and make overseas resources an important complement to its natural gas supply.

It is predicted that China's capability of natural gas supply will reach 370 billion m^3 by 2020, including the output of homemade natural gas (conventional natural gas, dissolved gas, and coal-bed gas) totaling 240 billion m^3. It plans to build four major natural gas supply bases in northwest China, southwest China, northeast China, and on the sea, and its output of natural gas is expected to increase steadily. The natural gas supply/demand gap will be met by means of the China-Central-Asia, China-Myanmar, and China-Russia natural gas pipelines, as well as the importing of liquefied natural gas, and the scale of supply by these means will increase steadily. It is predicted that China is capable of importing 130 billion m^3 of natural gas in this way by 2020. The domestic supply-demand gap will widen after 2020, and it will be necessary to import more natural gas, and overseas resources will become an important complement to China's supply of natural gas. China's resources and supply capability of natural gas in 2020 are shown in Table 4.19.

(5) Energetically develop nonconventional natural gas may help to lower China's external dependency on natural gas to a certain extent.

Under the positive influence of the large-scale shale gas development in the United States, China has attached increasing importance to its development of

Table 4.19 China's resources and supply capability of natural gas (unit: 100 million m^3)

Category		2020
Conventional gas	Accumulative new proven geological reserves	65,000
	Output of natural gas	1700
	Output of dissolved gas	100
Coal bed gas	Accumulative new proven geological reserves	11,000
	Output	150
Coal gas	Output	150
Shale gas	Output	300
Imported gas	Pipeline natural gas and LNG	1300
Total supply		3700

Table 4.20 The predicted output and importing of China's natural gas in 2020 under the scenario of the large-scale development of shale gas (unit: 100 million m^3)

Year	Demand for natural gas	China's output of natural gas			Import of natural gas	External dependence
		Conventional natural gas[a]	Coal bed gas	Shale gas		
2020	3150	1800	150	300	1330	41.3%

[a]*Including the output of tight gas (natural gas of sandstone formations with a permeability of less than 0.1 milli-darcy). Now China considers tight gas as conventional gas.*

shale gas. China has abundant resources of shale gas, and its recoverable land shale gas resources may reach 25.08 trillion m^3[37] (excluding the Qinghai-Xizang Plateau), ranking the first in the world. According to the *"12th Five-year Plan" on the Development of Shale Gas*, China plans to increase its shale gas output to 60–100 billion m^3 by 2020. At that time, thanks to the large-scale development of shale gas, China's external dependency on natural gas may drop by about 10 percentage points. The predicted output and importing of China's natural gas in 2020 under the scenario of the large-scale development of shale gas are shown in Table 4.20.

(6) Gas consumption for power generation is one of the key factors for the increase of natural gas demand in the future.

Improving the natural gas consumption structure is the basic principle for China's utilization of natural gas. In October 2012, China's National Development and Reform Commission issued the new *Policy on Utilization of Natural Gas*. Compared with the 2007 edition of the policy, the new policy gives greater support to gas power generation. Between 2010 and 2030, among the average annual growth rates of major uses of natural gas in China, the average annual growth rate of gas consumption for power generation is the highest (9.2%), while that of gas for urban use, industrial fuel, and chemical use will reach 8.8%, 7.2%, and 5.6%, respectively.

In 2015, the gas for power generation accounted for 14.6% of the total natural gas consumption. It is predicted that the proportion of gas consumption for power generation to the total gas consumption will reach about 18% at the end of the "13th Five-year Plan" period. In the medium and long term, the proportion of natural gas for power generation will go up steadily.

---

[7]The Ministry of Land and Resources, A Survey and Evaluation of the Potential of China's Shale Gas Resources and Favorable Area Optimization, March 2012.

Construction and comprehensive evaluation of non-fossil energy development scenarios

This chapter analyzes the possible development scale of major non-fossil energy power generation, such as hydraulic generation, nuclear power, wind energy, and solar energy. In this chapter, scenario analysis methods and overall power system optimization planning models are used to study the generation and transmission capacity expansion under different clean energy[1] power generation development scenarios. The investment, operating cost, as well as various economic and social benefits of each scenario are also studied, so as to provide references for making decisions in choosing the path toward achieving the goal of making non-fossil energy use account for 15% of China's total energy consumption.

5.1 THE MAIN PRINCIPLES OF SCENARIO ANALYSIS

Safety. The installed power capacity and cross-regional power transmission should be capable of meeting the system load demands, while maintaining a reasonable reserve level; the power sources should be operating in a coordinated way, meet the load demands, and track load changes in a timely manner, so as to ensure safe and stable operation of the power system. The system generation and transmission expansion plan shall meet the national and regional demands for the electric quantity balance and peak shaving balance demands.

Economy. The investment and operating cost of various power sources should be fully considered, and the external cost of pollutant emissions from fossil energy power generation should be considered on the basis of the investment and operating costs of power transmission lines. Moreover, the relationship of coordinated development between clean energy and electric power is studied, with the purpose of minimizing the total cost of electric power supply for China and the on the basic condition of meeting the bearing capacity of power consumers.

---

[1]The "clean energy" in this book refers to clean energy in a narrow sense. It covers nuclear energy and renewable energy, and its scope is the same as that of non-fossil energy. Power generation based on clean energy refers to power generation based on such non-fossil energy sources as hydro, nuclear, wind, solar and biomass energy sources, etc.

Non-Fossil Energy Development in China. https://doi.org/10.1016/B978-0-12-813106-0.00005-2

Cleanliness. Under the precondition of ensuring safe operation of the power system, strengthen power construction to expand the scope of a clean energy market, accelerate peak shaving power source construction to enhance the clean energy consumption capability, optimize scheduling of power generation units in system operations, make the best of the regulating capability of peak shaving power sources and the allocation capability of cross-regional power grids, maximize the development scale and consumption of clean energy sources such as hydropower, nuclear power, wind power, and solar energy, reduce the fossil energy consumption, environmental pollutants and greenhouse gas emissions of the power industry, and promote the green and low-carbon development of the power industry.

5.2 MAIN BOUNDARY CONDITIONS
5.2.1 PREDICTIONS ON ENERGY DEMANDS
5.2.1.1 Predictions from relevant domestic organizations

The Chinese Academy of Engineering analyzed the necessity and possibility of China's economic structural adjustment, in its Strategic Research on China's Medium and Long-term (2030 and 2050) Energy Development, energy consumption per unit of gross domestic product (GDP) of China as well as the possibilities of structural energy saving, industrial energy saving, architectural energy saving, transportation energy saving, and social consumption energy saving. The research results show that it is possible to "control the per-capita energy consumption of China significantly to less than that of developed countries at the same period." It proposes to support economic development with a relatively low-energy elasticity coefficient (less than 0.5, reduced with time) and to minimize the total energy consumption (especially fossil energy consumption) when China is on the road for achieving its third strategic goal, and it is possible for China to keep its total energy consumption to under 4 billion tons of standard coal by 2020 and to 4.5 billion tons of standard coal by 2030. Moreover, according to the research, achieving the strategic target is extremely challenging, but it is of great significance, symbolizing the actual effects of transforming the growth model, and is also the key to meeting the commitments of the Chinese government in coping with climate changes.[2]

In 2009, China's competent energy authority organized the formulation of the Scientific Development 2030—Research Report on China's National Energy Strategy, predicting China's development tendency of energy demands, drawing on the

---

[2]In November 2009, before the Copenhagen Climate Change Conference, the Chinese government announced that it would reduce its carbon dioxide (CO_2) emissions per unit of GDP by 40%–45% by 2020, compared with 2005 levels. In June 2015, before the Climate Change Conference held in Paris, China submitted its Intended Nationally Determined Contributions (INDC), pledging to peak its CO_2 emissions by 2030, reduce its CO_2 emissions per unit of GDP by 60%–65%, compared with the 2005 levels, and achieve the goal of making the proportion of non-fossil energy use account for about 20% of the total primary energy consumption.

changing rules of economic and social development and energy consumption of major developed countries in different development stages, and giving full consideration to China's economic and social development and characteristics of energy demand. Through contrastive analyses on major influencing factors and international experience, the report considers that, similar to the economic and social development process, from now to the middle of the 21st century, China's energy intensity per unit of GDP will show a downward tendency, and the total primary energy demand will show a rising trend, but have different characteristics in different periods.

Between 2010 and 2020, the Chinese government intensified and will continue to intensify its efforts for industrial structural adjustment, take strong measures to shut down outdated facilities, and work hard to give play to the role of technological advancement and energy price in regulating energy demands. However, the energy-intensive basic industries of energy and raw materials still form a large part and are growing rapidly, since China is at the stage of rapid development of industrialization and urbanization. Although the energy intensity will decline rapidly, the energy demand will continue to rise rapidly.

Between 2021 and 2030, the proportion of non-fossil energy use, to the total primary energy use of energy-intensive basic industries of energy and raw materials, will keep declining, while that of high-processing manufacturing industry, with lower energy consumption will keep rising, that of the tertiary industry will also grow rapidly, industrial structural adjustments will bring obvious energy-saving effects, and energy-saving technologies and alternative energy sources will develop rapidly. Therefore, energy intensity will continue to drop rapidly, and the total energy demand will rise steadily.

By using methods for both top-to-bottom and bottom-to-top predictions, the research report predicts energy demands for China and has obtained consistent results through two-way integration. The basic setting of the baseline plan is based on the economic and social development, the baseline plan does not only cover the industrial structural upgrade and the increase of energy efficiency, driven by economic and technological development, on the basis of current policies, but also considers the factor of strengthening policy on economic structural optimization and technological advance, under the dual pressures of shortages of energy supply and climate change, so as to effectively implement the policies on structural adjustment and energy saving and make significant progress in the transformation of the economic growth model. In this scenario, it is predicted that China's total energy demand will reach 4320 million tons of standard coal in 2020. Considering the uncertainty of economic and social development, on the basis of high and low estimates, it is predicted that China's total energy demand will be between 3.8 and 4.9 billion tons of standard coal by 2020, and between 5.5 and 6 billion tons of standard coal by 2030.

5.2.1.2 Predictions from international organizations

Among the research results of major international organizations, the World Energy Outlook 2011 of the International Energy Agency (IEA) predicts the baseline scenario of China's total energy demand of 4.78 billion tons of standard coal in

2020, with a minimum total energy demand of 4.55 billion tons of standard coal and a maximum total energy demand of 4.95 billion tons of standard coal; the International Energy Outlook 2011 of the US Energy Information Agency (EIA) predicts a baseline scenario of 4.67 billion tons of standard coal, within a scope of between 4.48 and 4.87 billion tons of standard coal.

5.2.1.3 Predictions of State Grid Energy Research Institute

By using the methods and models such as regression analysis and system, the State Grid Energy Research Institute predicts that China's total demand for primary energy will reach 4.8–5.1 billion tons of standard coal in 2020, and 5.6–5.9 billion tons of standard coal in 2030.

By this time, during the "13th Five-year Plan" period (2016–20), China is at the late stage of industrialization, and its industrialization and urbanization will continue to develop at a high rate. It is estimated that, by 2020, the average annual rate of GDP will reach about 6.5%, the GDP will reach RMB 93 trillion,[3] and the GDP per capita will reach RMB 66,000 (USD 10,600[4]).

Between 2010 and 2020, as the Chinese government intensifies its efforts for industrial structural adjustment and takes progressive measures to shut down outdated production facilities, China's energy intensity will drop noticeably. However, the energy-intensive basic industries of energy and raw materials still form a large part and will grow rapidly, since China is at the stage of rapid development of industrialization and urbanization, and the overall energy demand remains at a high growth rate.

Between 2021 and 2030, China will basically complete its industrialization process and continue to advance its urbanization process. It is predicted that China's GDP will grow at an average annual rate of 4%–5%. By 2030, China's GDP will reach RMB 138–151 trillion, and the per-capita GDP will reach RMB 95,000–104,000 (USD 15,000–17,000). During this period, the proportion of non-fossil energy use to the total primary energy use of energy-intensive basic industries of energy and raw materials in secondary industry will drop continuously, while that of the manufacturing industry, with relatively low energy consumption and high processing, will keep rising, and that of tertiary industry will also grow rapidly, while industrial structural adjustment will bring obvious energy-saving effects, and energy-saving technologies and alternative energy sources will develop rapidly. Therefore, the energy intensity will continue to drop rapidly, and the total energy demand will rise steadily.

Diversely, it is estimated that China's coal demand will reach its peak ranging from 4 to 4.5 billion tons before 2020, and enter a period with declining consumption after 2020. It is estimated that the petroleum demand will continue to grow rapidly before 2020, with an average annual increase of about 20 million tons, and after

---

[3]Constant price in 2015, the same below.
[4]As per RMB 6.23/USD 1, the average exchange rate for RMB to USD in 2015, the same below.

2030, the consumption growth will gradually slowdown. It is estimated that the natural gas demand will grow rapidly before 2030, and the proportion of natural gas demand to the total energy demand will continue to rise.

5.2.1.4 Comprehensive analysis on energy demands in the future

The detailed results of the above predictions from the above domestic and overseas research organizations on China's energy demands are shown in Table 5.1. It is estimated that by 2020 China's total primary energy demand is very likely to be from between 4.5 and 5.1 billion tons of standard coal, and the per-capita energy consumption will be 3.2–3.6 tons of standard coal, slightly higher than the world's current per-capita energy consumption level; between 2011 and 2020, the average annual growth rate will be 3.4%–4.6% (it was 9.4% between 2000 and 2010), with an average annual growth of 130–185 million tons of standard coal (it was about 190 million tons of standard coal between 2000 and 2010). It is predicted that the total energy consumption will be between 5.5 and 6 billion tons of standard coal in 2030, and the per-capita energy consumption will be between 3.8 and 4.1 tons of standard coal; during the period from 2021 to 2030, China's average growth rate will be between 1.6% and 2.0%, and the average annual growth will be about 90–100 million tons of standard coal. The following analyses are made under the high, medium, and low boundary conditions assuming the total energy demand is 5.1, 4.8, and 4.5 billion tons of standard coal for 2020, and 5.5, 5.8, and 6 billion tons of standard coal for 2030.

5.2.2 POWER DEMAND

The State Grid Energy Research Institute made predictions for the medium- and long-term power demands of China under the high and low plans.

Table 5.1 Prediction on China's energy demands in 2020 and 2030 (unit: 100 million tons standard coal)

Organization	Scenario	2020	2030
National Energy Administration	Baseline	43.2	57
	Low	38.0	55
	High	49.0	6
International Energy Agency (IEA)	Baseline	47.8	–
	Low	45.5	–
	High	49.5	–
US Energy Information Administration (EIA)	Baseline	46.7	–
	Low economic growth	44.8	
	High economic growth	48.7	–
State Grid Energy Research Institute	Baseline	48.0	58
	Slow transformation	51.0	55
	Fast transformation	45.0	6

5.2.2.1 Analysis and prediction of the economic situation

Since the adoption of the policy of reform and opening-up, China has made impressive achievements in its economic and social development. Between 1978 and 2015, China's GDP grew at an average annual rate of 9.7%; in 2015, its economic aggregate reached USD 10.9 trillion (current price), ranking the second in the world; the annual average per-capita GDP grew at a rate of 8.6%, and reached USD 7924/person in 2015. During the 13th "Five-year Plan" period, China has entered a decisive stage of completing the building of a moderately prosperous society in all respects. As is stated in the report delivered at the 18th National Congress of the Communist Party of China (CPC), China will double its 2010 GDP and per capita income for both urban and rural residents, and complete the building of a moderately prosperous society in all respects by 2020. To achieve this target, its economy needs to grow at an average annual rate of at least 6.5% during the 13th Five-year Plan period. Meanwhile, it also indicates that China will enter a new phase of economic development in which quality enhancement rather than quantitative growth is pursued. During the period between 2021 and 2030, China will further adjust and optimize its economic structure, encourage the development of the service sector and support the development of emerging industries, lower the level of labor supply, raise the quality of labor, and steadily raise per capita labor productivity. By considering such factors as international political and economic trends, domestic industrial structural adjustment, labor supply, and technological advancement, the predicted economic growth under the two scenarios between 2021 and 2030 is shown further.

Low growth scenario. The industrialization process basically draws to an end, infrastructure keeps improving, investment growth further slows down, and consumption reaches a relatively high level; there is no severe economic turmoil in the world, international and domestic trade will tend to balance; the economy has entered a stage of organic growth innovation-driven development, and the contribution of technological advances to the economy has increased significantly. The measures for market-oriented reform have been in place, and a sound market-oriented operation mechanism has been basically established. It is predicted that China's economy will grow at an average annual rate of about 4% between 2021 and 2030, the GDP will rise to about 22.2 trillion USD (price and exchange rate of 2015, the same below) by 2030, and the per-capita GDP will reach USD 15,000/person, 1.9 times of that in 2015.

High growth scenario. The international economic environment keeps steady, strategies such as the "Belt and Road" initiative and the internationalization of RMB have delivered tangible results, the international operation level of enterprises has significantly increased, China's initiative in taking part in international governance has further increased; China has made a greater breakthrough in technological innovation, reached or approached the world leading levels in the fields of smart manufacturing, Internet Plus, new energy, and new materials, and the driving effects of technological innovation on economic growth have been further enhanced; the market-oriented economic mechanism and social security mechanism have been improved, and residents are more willing to increase their income and spend money.

It is estimated that the average annual economic growth rate between 2021 and 2030 will reach about 5%. By 2030, China's GDP will reach about USD 24.2 trillion, and its per-capita GDP will reach USD 17,000/person, 2.1 times of that in 2015.

5.2.2.2 Prediction of total power demand

On the basis of predictions of the economic development and the total energy consumption of China, the analytical results of some advisory agencies and the forecasts regarding the power demands of different provinces, and the growth trend of the actual power consumption between 2000 and 2015, through the comprehensive analyses on such factors as China's industrialization and urbanization, the increase of the level of electrification, the adjustment of the tertiary industrial structure, the adjustment of industrial distribution, population growth and the world economic situation, etc., it is estimated that in the next 15 years China's power demands are likely to continue to rise significantly, but the growth of power consumption will drop markedly compared with that of the past 15 years. The predictions on the two economic growth scenarios are shown further.

Scenario 1 (low growth scenario). It is estimated that China's power consumption will reach 7.0 trillion kWh by 2020, with an average annual growth rate of 4.3% during the "13th Five-year Plan" period, and the power elasticity coefficient will reach 0.66. By 2020, the power consumption per unit of GDP will be 690 kWh/10,000 yuan (820 kWh/10,000 yuan in 2015), and the level of electrification will reach 2% (23% in 2015). Due to structural adjustments and improvements in living standards, the power consumption of tertiary industry and households will grow rapidly, and the growth of power loads will exceed that of power supply. It is predicted that the power load will grow at an average annual rate of 5.0% during the "13th Five-year Plan" period, about 0.7 percentage points higher than that of power consumption, and power loads will reach 1110 GW by 2020. Between 2020 and 2030, the average annual growth rate of China's power consumption and power loads will be 2.5% and 2.7%, respectively, and reach 9.0 trillion kWh and 1.45 billion kW by 2030, respectively.

Scenario 2 (high growth scenario). It is estimated that China's power consumption will reach 7.13 trillion kWh by 2020, with an average growth rate of 5.1% during the 13th Five-year Plan period and a power elasticity coefficient of 0.78. By 2020, the power consumption per unit of GDP will be 760 kWh/RMB 10,000 yuan, and the level of electrification will reach 28%. It is estimated that in the next 10 years the average annual growth rate of power load will be 5.9%, about 0.29 percentage points higher than the growth rate of power consumption. By 2020, China's power loads will reach 1.16 billion kW. Between 2020 and 2030, the average annual growth of China's power consumption and power loads will reach 2030 10.0 trillion kWh and 1.63 billion kW, respectively, at a growth rate of 2.7% and 3.4%, respectively,.

Through analyses on the two scenarios, it is estimated that China's total electricity demand will reach 7.0–7.3 trillion kWh by 2020, and 9.0–10.0 trillion kWh by 2030, see Table 5.2 for the prediction results of power demands.

Table 5.2 Predictions on power demands between 2010 and 2030

Predictions on energy power demand		Demand			Growth rate		
		2015	2020	2030	2011–15 (%)	2016–20 (%)	2021–30 (%)
Power consumption (trillion kWh)	Low	5.55	7.0	9.0	7. 4	4.3	2.5
	High	5.55	7. 3	10.0	8. 6	5.1	3.2
Maximum load (100 GW)	Low	8.9	11.1	14.5	8. 0	5.0	2.7
	High	8.9	11.6	16.3	8 8	5.9	3.4

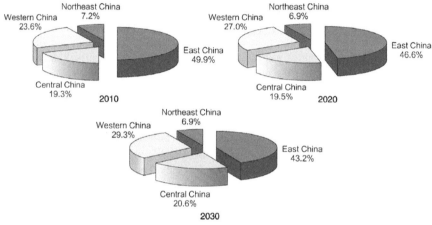

FIG. 5.1

The distribution of power demands between 2010 and 2030. Note: Eastern and central China in the diagram above include 10 provinces and autonomous regions such as Beijing, Tianjin, Hebei, Shandong, Shanghai, Jiangsu, Zhejiang, Fujian, Guangdong, and Hainan, central China includes six provinces such as Shanxi, Henan, Hubei, Hunan, Anhui, and Jiangxi, western China includes 12 provinces (autonomous regions, municipalities) such as Nei Mongol, Sichuan, Chongqing, Shaanxi, Gansu, Qinghai, Ningxia, Xinjiang, Xizang, Guangxi, Yunnan, and Guizhou, northeast China includes Heilongjiang, Jilin, and Liaoning.

As is seen from distribution characteristics, considering such factors as economic structure, industrial structure, structural adjustment, and population distribution of China, the proportion of electricity use in western China will increase in the future. Nevertheless, the eastern and central regions will continue to be the power load center of China for a long time to come, due to the large number of bases, as is shown Fig. 5.1. In 2015, the power consumption of the 16 provinces and municipalities in eastern and central China[5] accounts for 68.4% of that of the whole of China, and will reach 66.1% by 2020.

---

[5] The 16 provinces in eastern and central China are the 10 provinces (municipalities) in eastern China and 6 provinces in central China.

5.2.3 NON-FOSSIL ENERGY DEVELOPMENT SCALE

At the United Nations Climate Change Conference held in December 2015, nearly 200 contracting states to the United Nations Framework Convention on Climate Change agreed, by consensus, to the Paris Agreement, making arrangements for a global campaign against climate change after 2020 and limiting the global temperature increase to 2°C. The Chinese government pledged at the conference to peak its carbon emissions around 2030. To achieve the goal, the Chinese government will adopt the essential energy strategies of adjusting energy structure and accelerating the non-fossil energy development in the future.

The development of nuclear power is an important choice of the world's major developed countries for adjusting their energy structure, and also an effective method for China in realizing its goal of energy structural adjustment, but the development of nuclear power in China is restricted by nuclear power security, construction cycle, supply of equipment, etc. Thus, it is uncertain whether China can achieve its maximum development scale in 2020. China boasts rich hydroenergy resources and mature technologies, nevertheless, since the resources are less developed, energetically developing hydropower is a necessary and realistic choice for China to achieve its energy structural adjustment goal. Hydropower development is restricted by such factors as immigration, environmental protection, construction cycle, etc., and its maximum development scale for 2020 is also limited. Actively developing power generation based on renewable energy sources including wind energy and solar energy is a strategic measure taken by major developed countries in the world to cope with climate changes, safeguard energy security, and protect the ecological environment. Nevertheless, restricted by natural conditions, the utilization time of equipment for power generation, based on renewable energy sources, including wind energy and solar energy equipment, is much shorter than that for power generation based on conventional energy, and the energy density[6] of such renewable energy sources is very low. Meanwhile, wind and solar power output is random and intermittent, and its consumption capability is restricted by structure, distribution, and scale of the electric power system. Before 2020, the development scale, distribution, and consumption market of wind power and solar power shall be optimized on the basis of the overall development plan of the electric power system. Therefore, there is a possible ceiling of the development scale of hydropower, nuclear power, and wind power in 2020. In 2030, if the preliminary work goes smoothly and the construction cycle is guaranteed, the development scale of clean energy power generation will depend mainly on resource reserves, development conditions, and the consumption capacity of the electric power system.

---

[6]Generally, energy density is the amount of energy stored in a given system or region of space per unit volume or mass. In this book, it refers in particular, to annual electricity generated from unit installed capacity.

5.2.3.1 Hydropower

In 2015, China's installed conventional hydropower capacity was 290 GW. In accordance with the "13th Five-year Plan" on electric power development and the relevant plans on hydropower construction in southwest China,[7] between 2016 and 2020, it is planned to increase the installed hydropower capacity in Sichuan and Yunnan by 20 GW and 20 kW, respectively. If the above plans have been implemented, China's hydropower development scale will exceed 330 GW by 2020; if new hydropower stations in other provinces and small hydropower stations are included,[8] China's hydropower development scale is likely to reach 350 GW by 2020.[9] In 2030, the hydropower resources on the upper reach of the Jinsha River, the upper reach of the Yalong River, the Nu River and the Yarlung Zangbo River in southwest China will be further developed, the installed hydropower capacity of China is expected to reach 400–440 GW, and China's hydropower development and utilization ratio is expected to reach about 80%.[10]

5.2.3.2 Nuclear power

At the end of 2015, China's installed nuclear power capacity reached 27.17 kW, the installed capacity of underconstruction nuclear power projects was about 27 GW, and the total installed nuclear power capacity is expected to reach about 54 GW. China plans to build nuclear power stations with an installed capacity of 20–30 GW between 2016 and 2020, and China's installed nuclear power capacity will reach nearly 60 GW at that time. Affected by Japan's nuclear power disaster, the approval process of China's nuclear power stations has slowed down over the past 2 years, and the nuclear power development scale is likely to fail to meet the set target by 2020. According to the "13th Five-year Plan" on electric power development, the installed capacity of operating nuclear power stations may reach 50–60 GW by 2020, accounting for about 3% of the total installed capacity of the country.[11] Against the backdrop of energy structural adjustment and transformation and the nonrenewability of fossil energy, the development of nuclear power is essential to ensuring China's energy supply security in the medium and long term, improving the energy structure and achieving the targets of energy conservation and emission reduction. It is predicted that between 2020 and 2030 China will continue to build nuclear power stations on a large scale, and the its installed nuclear power capacity is expected to reach 150 GW by 2030.

---

[7]The Grid Development Plan of the State Grid Corporation of China for the 12th Five-year Plan Period, The Research Report of Southern China Power Grid on the Electric Power Industrial Development during the 12th Five-year Plan and on a medium and long term.

[8]As of the end of 2008, China's installed capacity of small hydropower stations was 51 GW. According to the medium- and long-term renewable energy planning, it is expected to reach 75 GW by 2020.

[9]The "12th Five-year Plan on Renewable Energy Development."

[10]Hydropower development and utilization ratio refers to the ratio between the installed capacity of built hydropower stations and technically available hydropower resources.

[11]It is predicted that China's total installed power capacity will reach 1.9–20 billion kW in 2020.

5.2.3.3 Wind power

China's installed grid-connected wind power capacity was 130 GW at the end of 2015. On the basis of the development potential of wind energy resources, as well as the conditions of wind energy resources, wind power farm sites, engineering geology, transportation, construction, installation, and project investment, China's wind power development potential is expected to exceed 300 kW by 2020, and the development potential is estimated on the basis of considering that the currently available resources can be economically developed and utilized. In the "13th Five-year Plan" on electric power development and the "13th Five-year Plan" on wind power development, the Chinese government aims to raise the wind power development scale to over 210 GW by 2020. Wind power features short construction cycles and adequate capacity of equipment manufacturing, and wind power farm construction capability will not restrict the development of wind power. On the basis of making proper research and planning on wind power grid connection and market consumption, it is possible for the wind power development scale to reach 200–230 billion kW by 2020.

5.2.3.4 Solar power

At the end of 2015, China's grid-connected installed solar power capacity was 41.58 GW. China is endowed with rich solar energy resources, has a tremendous potential for solar development, and its installed capacity is mainly restricted by the economical efficiency of power generation and the consumption capability of the power grid. China has released the "13th Five-year Plan" on electric power development and the "13th Five-year Plan" on solar energy development, setting a goal of making its development scale of solar power exceed 110 GW by 2020, including distributed PV of over 60 GW and photothermal power of 5 GW. Solar generation features construction cycles and adequate equipment manufacturing capacity, and the construction capability will not restrict the solar energy development. On the basis of making proper plans on grid connection, transmission, and consumption, China's solar generation development scale is expected to reach 100–130 GW by 2020.

Thanks to technical advances and further improvements of the economical efficiency, such renewable energy sources as wind power and solar power will become important energy supplies, and their development scale will continue to expand. It is estimated that by 2030 the installed capacity of wind power will reach 400–500 GW, and that of solar power is expected to reach 300–350 GW.

In the final analysis, the potential installed capacity of China's major clean energy sources are shown further: by 2020, China's installed capacity of hydropower, nuclear power, wind power, and solar power is expected to reach 300–350, 60, 200–230, and 100–130 GW, respectively; by 2030, China's installed capacity of hydropower, nuclear power, wind power, and solar power is expected to reach 400–440, 130–150, 400–500, and 300–350 GW, respectively.

5.2.4 **TOTAL SUPPLY OF NON-FOSSIL ENERGY**

As mentioned before, it is estimated that China's primary energy demand will reach 4.5–5.1 billion tons of standard coal by 2020. In accordance with the target of non-fossil energy use accounting for 15% of the energy consumption mix, the total non-fossil energy supply will reach 680–770 million tons of standard coal. According to relevant research results of Scientific Development 2030—Research Report on China's National Energy Strategy, by 2020, the total scale of non-fossil energy for purposes other than power generation, such as solar energy heat supply, biomass energy gas supply, biomass energy fuel, and geothermal utilization, is expected to reach about 130 million tons of standard coal. If such noncommercial energy sources as biomass gas supply and solar energy heat utilization are not considered, the total scale of non-fossil energy for purposes other than power generation will reach about 40 million tons of standard coal.[12] This book uses research results such as boundary conditions for computation, and therefore, the total scale of non-fossil energy for power generation needs to reach 550–640 million tons of standard coal, accounting for over 80% of the total non-fossil energy consumption. By 2030, China's total primary energy demand will reach 5.5–6 billion tons of standard coal, and the total supply of non-fossil energy is expected to reach 1.1–1.3 billion tons of standard coal, accounting for over 20% of the total energy supply.

5.3 **DEVELOPMENT SCENARIO DESIGN**

5.3.1 **GENERAL IDEA**

The scenario design shall be typical and representative, the relevant national energy development plans shall be referred to, and the actual development level and various uncertainties shall be fully considered. In this book, four types of typical scenarios are analyzed: (1) Expected development scenario, under which the expected development scale of renewable energy sources such as hydropower, nuclear power, and wind power has been achieved. The scenario is also the baseline for comparison with other scenarios; (2) Hydropower development slowdown scenario, mainly considers the uncertainties of hydropower operation time sequence caused by the long cycle of hydropower construction, resettlement, and environmental protection and expects that the construction and operation scale is likely to fall significantly; (3) Nuclear power development slowdown scenario. This mainly proposes that the Chinese government shall be more prudent in nuclear power development for the purpose of ensuring nuclear power security. The scenario presents the non-fossil energy development path of nuclear power under the condition of prudential development;

---

[12]According to the Scientific Development 2030—Research Report on China's National Energy Strategy of National Energy Administration, the total installed capacity of non-fossil energy for purposes other than power generation is 130 million tons of standard coal, specifically, 35 million tons for biomass gas supply, 57 million tons for heat utilization of solar energy, 10 million tons for geothermal heat utilization, 14 million tons for biomass solid particles, and 14 million tons for biological liquid fuel.

(4) Power demand slowdown scenario. This studies non-fossil energy development under the condition of market demand slowdown in the future, with a low plan on power demand as the boundary.

In the scenario design, first of all, the power grid zoning plan and the power source development plan as input conditions shall be defined, and the to-be-selected type of power sources and the maximum power flow shall be optimized. The main ideas of this chapter are shown further.

5.3.1.1 Power grid zoning plan

By power sending and power receiving ends, China's power grids are divided into western Nei Mongol, northeast China (including eastern Nei Mongol), Shanxi, northwest China, receiving ends in eastern and central China (including Beijing, Tianjin, Hebei, and Shandong in North China, as well as the 13 provinces and municipalities in eastern China and central China, the same below), Chuan-Yu-Zang (Sichuan, Chongqing, and Xizang), and southern China. The topological relation between the above major regions is shown in Fig. 5.2.

5.3.1.2 Plan on power source construction as input condition

By referring to national plans such as the Strategic Action Plan on Energy Development (2014–2020), the "13th Five-year Plan" on energy development and the "13th Five-year Plan" on electric power development, the process of power source structural optimization, hydropower, nuclear power, wind power, solar power, and biomass power are defined as power sources, with their development scale for 2020

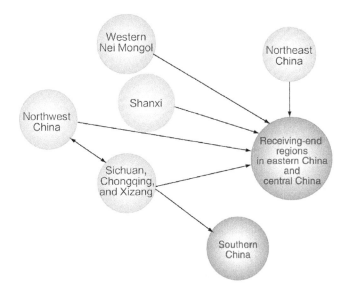

FIG. 5.2

The topological relationship between major regions in China.

in line with China's national plans; for the development distribution of the energy sources, resources, energy and renewable energy planning, as well as preliminary work of the seven designated power grid areas shall be referred to, and calculations shall be made according to the designated scale and time sequence as input boundary conditions; the installed capacity of coal-fired thermal power, which is also considered a fixed power source, is mainly related to the growth of regional heat supply load.

5.3.1.3 Optimized and alternative variables

The installed capacity, time sequence, and distribution of conventional coal power, gas power, and pumped storage, which are optimized and alternative power sources, are determined by the software for overall optimization planning of the electric power system (GESP-IV). The methods for analysis on wind power development scale and consumption capability are shown in Section 3.2, and the boundary conditions of techno-economic parameters for power source optimization are shown in Appendix II.

5.3.1.3.1 Conventional coal power

Underconstruction and approved coal power projects shall be launched in accordance with their time sequence. Optimized and alternative coal power shall be simplified into 0.6 and 1 million kW units in the calculation of GESP-IV,[13] its production scale and progress are optimized and determined by GESP-IV, and the maximum production capacity of coal-fired units in different planning periods shall be set in accordance with energy resources and power source development conditions.

5.3.1.3.2 Pumped storage

Pumped storage projects that are under construction, that have been approved, or that have obtained preliminary permits shall be launched in the planned year. In addition, in the areas that have conditions for site selection for construction of pumped storage power stations, alternative pumped storage projects with a certain capacity shall be designed, and the production schedule and scale of such projects shall be optimized and determined by GESP-IV. The maximum capacity of alternative projects shall be determined on the basis of the total capacity of the alternative sites of all pumped storage power stations.

5.3.1.3.3 Gas power station

In spite of the high fuel cost, fuel gas power generation has prominent advantages of low maximum output, fast creep speed, and good regulation performance. It is a peak shaving power source with a good system. In the research, single-cycle and

---

[13]According to the Guidance Catalogue for Adjustment of Industrial Structure (2014) released by the National Development and Reform Commission, the development of conventional coal-fired thermal power units with a unit capacity of 300,000 kW and below is restricted.

combined-cycle gas power stations are taken as alternative power source for optimization, respectively. The type, production schedule, and scale of gas power stations shall be optimized and selected by the system on the principle of the economical efficiency.

5.3.1.4 Cross-regional power grid expansion scale (power flow)

In the overall optimization and planning model of the electric power system, cross-regional power flow also needs to be optimized. In this book, with the purpose of minimizing the total cost of electric power supply, China's power source planning and power flow-scale expansion are optimized, the cross-regional switching scale is determined on the principle of maximizing the cost of electric power supply, and the type, scale, and distribution of installed power capacity of different regions are determined.

5.3.2 FOUR TYPES OF SCENARIOS

5.3.2.1 Scenario 1: Expected development scenario

Scenario setting—the development scale of hydropower, nuclear power, wind power, and solar power has met the expectation.

Under boundary conditions (the development scale of hydropower, nuclear power, wind power, and solar power meets the expectations), the power flow and power source structure in correspondence with scenario 1, including the construction scale of corresponding coal power, gas power, and pumped storage, are studied. Moreover, reasonable consumption markets and ways of consuming renewable energy, as well as the typical operation mode of the electric power system, are analyzed.

5.3.2.1.1 Boundary conditions

According to the high demand scenario (China's power consumption is 7.3 trillion kWh in 2020), China's installed generation capacity in 2020 is shown further: hydropower (350 GW), nuclear power (58 GW), wind power (200 GW), solar power (100 GW), biomass power (17 GW), the capacity in 2030 is shown further: hydropower (430 GW), nuclear power (130 GW), wind power (400 GW), solar power (300 GW), and biomass power (30 GW). UHV AC/DC power transmission technologies are mainly used for cross-regional grid interconnection and expansion, and relevant parameters related to UHV AC/DC power transmission are used in grid investment and operation costs.

5.3.2.1.2 Plan for installed clean energy capacity

5.3.2.1.2.1 Hydropower. According to China's hydropower construction and the "13th Five-year Plan" on electric power development, it is estimated that China's installed hydropower capacity will reach 345 GW by 2020 and 430 GW by 2030. In the future, as is seen from the power grid areas, new hydropower stations will mainly be built in Sichuan, Chongqing, and Xizang in southwest China and the

Table 5.3 Arrangements of installed hydropower capacity (operating) in the power grid areas (unit: 10,000 kW)

No.	Region	2015	2020	2030
1	Receiving ends in eastern and central China	7769	8100	8525
2	Northeast China	654	800	853
3	Northwest China	2877	3463	4342
4	Sichuan, Chongqing, and Xizang	7744	10,049	15,043
5	Shanxi	124	150	222
6	Western Nei Mongol	86	100	175
7	Southern China	10,412	11,861	14,000
Total		29,666	34,523	43,160

northwest China power grid (mainly Qinghai). The arrangements of installed hydropower capacity of the power grid areas are shown in Table 5.3.

5.3.2.1.2.2 Nuclear power. Since the Japanese Fukushima Daiichi nuclear disaster, the Chinese government has suspended the approval of nuclear power projects for over 2 years. According to the "13th Five-year Plan" on electric power development, it is estimated that installed nuclear power capacity is likely to reach about 58 GW by 2020. The newly installed nuclear power stations will be mainly located in coastal areas in eastern China where the power loads grow rapidly and energy resources are lacking. Between 2021 and 2030, China will begin to develop nuclear power on a large scale in its energy-hungry inland provinces. It is predicted that the installed nuclear power capacity of China will increase by about 72 GW, with a total installed capacity of 130 GW by 2030. The arrangements of regional installed nuclear power capacity are shown in Table 5.4.

5.3.2.1.2.3 Wind power. In accordance with the distribution characteristics of China's wind energy resources, the Chinese government shall place equal emphasis on centralized development and distributed development of wind power. To make

Table 5.4 Arrangements of installed nuclear power capacity (operating) in power grid areas (unit: 10,000 kW)

No.	Region	2015	2020	2030
1	Receiving ends in eastern and central China	1414	2940	8167
2	Northeast China	300	400	900
3	Northwest China	0	0	0
4	Sichuan, Chongqing, and Xizang	0	0	0
5	Shanxi	0	0	0
6	Western Nei Mongol	0	0	0
7	Southern China	1003	2484	3988
Total		2717	5824	13,055

the best of clean electric power, the development of the wind resources in "Three Norths" regions, where large wind power bases are located, shall be given priority, since the areas have abundant wind energy resources and favorable development conditions. In addition, the advantages of the large power market of the eastern and central regions of China shall be fully leveraged to develop distributed wind energy resources and simulate local consumption according to local conditions. It is estimated that the installed wind power capacity will reach 200 GW by 2020 and 400 GW by 2030. The estimated installed wind power capacity of power grid areas, which is based on regional planning on wind power, is shown in Table 5.5.

5.3.2.1.2.4 Solar power. It is estimated that China's total installed solar power capacity will reach 100 GW by 2020, of which, 50 GW will be developed in a centralized way, and the remaining 50 GW will be developed in a distributed way; it is estimated that China's total installed solar power capacity will reach 300 GW by 2030, of which 180 GW will be developed in a centralized way, and the remaining 120 GW will be developed in a distributed way. In recent years, China has promoted a rooftop PV system in economically developed regions such as the Yangtze River Delta, the Pearl River Delta, and the Circum-Bohai Sea region and has built large grid-connected PV power stations of a certain scale in northwest China. After 2020, China's solar power development will rely mainly on the large grid-connected PV power stations in the desert, Gobi, and wastelands in northwest China. The arrangements of the estimated installed solar PV capacity in power grid areas are shown in Table 5.6, on the basis of regional solar generation planning.

5.3.2.1.2.5 Biomass power. China's installed biomass power capacity was 13 GW in 2015 and is expected to reach 17 GW by 2020. Considering that the biomass power shall be developed conditionally and in a limited way in China, it is predicted that the installed biomass power capacity will reach 30 GW and remain stable by 2030. The arrangements of installed biomass power capacity in power grid areas are shown in Table 5.7.

Table 5.5 Arrangements of installed wind capacity (operating) in power grid areas (unit: 10,000 kW)

No.	Region	2015	2020	2030
1	Receiving ends in eastern and central China	3121	5500	10,000
2	Northeast China	2466	3400	5800
3	Northwest China	3926	5400	10,000
4	Sichuan, Chongqing, and Xizang	84	700	1500
5	Shanxi	669	800	2000
6	Western Nei Mongol	1545	2400	5200
7	Southern China	1019	1800	5500
Total		12,830	20,000	40,000

Table 5.6 Arrangements of installed solar PV capacity (operating) in power grid areas (unit: 10,000 kW)

No.	Region	2015	2020	2030
1	Receiving ends in eastern and central China	1265	3700	9500
2	Northeast China	84	300	1000
3	Northwest China	2083	4000	10,000
4	Sichuan, Chongqing, and Xizang	53	400	2500
5	Shanxi	113	20	1000
6	Western Nei Mongol	412	60	2000
7	Southern China	148	80	4000
Total		2100	500	30,000

Table 5.7 The arrangements of installed biomass power capacity in power grid areas (unit: 10,000 kW)

No.	Region	2015	2020	2030
1	Receiving ends in eastern and central China	800	1000	1400
2	Northeast China	120	200	300
3	Northwest China	50	100	300
4	Sichuan, Chongqing, and Xizang	50	80	170
5	Shanxi	10	30	80
6	Western Nei Mongol	20	50	100
7	Southern China	250	240	650
Total		1300	1700	3000

5.3.2.2 Scenario 2: Hydropower slowdown scenario

Scenario setting—By 2030, the hydropower development scale will reduce by about 40 GW, and the wind power and solar power scale increase correspondingly so as to meet the non-fossil energy gap caused by hydropower slowdown.

Considering such factors as long hydropower construction cycles, resettlement, and environmental protection, as well as possible uncertainties, under scenario 2, if the development scale of nuclear power meets expectations (130 GW, 2030), the hydropower development scale will drop (400 GW, 2030), but the development scale of wind power and solar power will increase correspondingly. An analysis is conducted on the impact on the electric power system planning and operation, and a contrast is made between scenario 1 and scenario 2 to analyze the economical efficiency of scenario 2.

5.3.2.2.1 Boundary conditions

According to the high demand scenario (China's power consumption is 7.3 trillion kWh in 2020 and 10.0 trillion kWh in 2030), China's installed generation capacity in 2020 is shown further: hydropower (330 GW), nuclear power (58 GW), wind power

(250 GW), solar power (150 GW), and biomass power (17 GW) and in 2030 is shown further: hydropower (390 GW), nuclear power (130 GW), wind power (400 GW), solar power (300 GW), and biomass power (30 GW).

The plan for the distribution of adjusted installed clean energy capacity.

5.3.2.2.1.1 The plan on installed capacity in the case of hydropower slowdown.

China's installed hydropower capacity is expected to reach 330 GW by 2020 and 390 GW by 2030, see Table 5.8. Specifically, the installed hydropower capacity at the receiving ends in eastern and central China falls by 3 GW, in Sichuan, Chongqing and Xizang falls by 24 GW, in northwest China falls by 3.8 GW, and in southern China falls by 7 GW, while that in other power grid areas remains basically unchanged.

5.3.2.2.1.2 The plan on accelerated installed wind power capacity.

China's installed wind power capacity is expected to reach 220 GW by 2020 and 450 GW by 2030, see Table 5.9. On the principle of placing equal emphasis on centralized development and distributed development, compared with scenario 1, scenario 2 sees an increase in installed wind power capacity of 10 GW in the receiving ends in eastern and central China, an increase of 10 GW in northwest China, an increase of

Table 5.8 The distribution of China's installed hydropower capacity by 2030 (390 GW) (unit: 10,000 kW)

No.	Region	2015	2020	2030
1	Receiving ends in eastern and central China	7769	8000	8225
2	Northeast China	654	700	800
3	Northwest China	2877	3263	3962
4	Sichuan, Chongqing, and Xizang	7744	9549	12,583
5	Shanxi	124	150	200
6	Western Nei Mongol	86	100	150
7	Southern China	10,412	11,461	13,300
Total		29,666	33,223	39,220

Table 5.9 The distribution of the installed wind capacity (450 GW) by 2030 (unit: 10,000 kW)

No.	Region	2015	2020	2030
1	Receiving ends in eastern and central China	3121	6200	11,000
2	Northeast China	2466	3500	6100
3	Northwest China	3926	5800	11,000
4	Sichuan, Chongqing, and Xizang	84	800	2000
5	Shanxi	669	900	2400
6	Western Nei Mongol	1545	2600	6000
7	southern China	1019	2200	6500
Total		12,830	22,000	45,000

10 GW in southern China, an increase of 0.8 GW in western Nei Mongol, and further increases in other regions.

5.3.2.2.1.3 The plan on accelerated installed solar power capacity. China's installed solar power capacity is expected to reach 120 and 350 GW by 2020 and 2030, see Table 5.10. Compared with scenario 1, under which there is a total solar power capacity of 300 GW, scenario 2 sees increases of installed solar power capacity in all the regions.

5.3.2.3 Scenario 3: Nuclear power slowdown scenario

Scenario setting—in 2030, the nuclear power development scale will be reduced by about 30 GW, and the development scale of wind power and solar power will increase so as to meet the gap of non-fossil energy supply caused by nuclear power slowdown.

Considering that China's nuclear power development may further slowdown and there are uncertainties in the operation of nuclear power stations, the installed nuclear power capacity for 2020 is adjusted from 58.2 to 51.74 GW, down by about 6.5 GW.

By 2030, the installed nuclear power capacity will be reduced from 130 to 100 GW, down by 30 GW, and the installed capacity of wind power, solar power, and biomass power will reach 400 GW, 400 kW and 30 GW, respectively.

5.3.2.3.1 Boundary conditions

According to the high demand scenario (China's power consumption is 7.3 trillion kWh in 2020 and 10.0 trillion kWh in 2030), China's installed generation capacity in 2020 is shown further: hydropower (345 GW), nuclear power (52 GW), wind power (200 GW), solar power (150 GW), and biomass power (17 GW) and in 2030 is shown further: hydropower (430 GW), nuclear power (100 GW), wind power (400 GW), solar power (400 GW), and biomass power (30 GW).

The plan for the distribution of adjusted installed clean energy capacity.

Table 5.10 The distribution of installed solar PV capacity (350 million) by 2030 (unit: 10,000 kW)

No.	Region	2015	2020	2030
1	Receiving ends in eastern and central China	1265	4400	11,000
2	Northeast China	84	500	1300
3	Northwest China	2083	4500	10,700
4	Sichuan, Chongqing, and Xizang	53	500	3400
5	Shanxi	113	300	1400
6	Western Nei Mongol	412	700	2400
7	Southern China	148	1100	4800
Total		4158	12,000	35,000

5.3.2.3.1.1 The plan for installed nuclear power capacity in the case of slowdown. According to the distribution of nuclear power projects that are under construction, have been approved, or have obtained preliminary permits, China's installed nuclear power capacity will be 52 GW by 2020, and 100 kW by 2030, see Table 5.11. Specifically, the installed nuclear power capacity will fall by 21.60 GW at the receiving ends in eastern and central China, fall by 8 GW in southern China, and fall by 1 GW in northeast China, and the installed nuclear power capacity will remain unchanged in other power grid areas.

5.3.2.3.1.2 The plan on acceleration of installed solar power capacity (150 GW by 2020 and 400 GW by 2030). By 2020, China's installed solar power capacity will increase to 150 GW. Specifically, it will increase by 15 GW at the receiving ends in eastern and central China, 10 GW in northwest China, 0.9 GW in southern China, 5 GW in northeast China, and increase to different extent in other regions. By 2030, China's installed solar power capacity will reach 400 GW, see Table 5.12.

Table 5.11 The distribution of installed nuclear capacity (100 GW) by 2030 (unit: 10,000 kW)

No.	Region	2015	2020	2030
1	Receiving ends in eastern and central China	1414	2640	6000
2	Northeast China	300	400	800
3	Northwest China	0	0	0
4	Sichuan, Chongqing, and Xizang	0	0	0
5	Shanxi	0	0	0
6	Western Nei Mongol	0	0	0
7	Southern China	1003	2134	3200
Total		2717	5174	10,000

Table 5.12 The distribution of installed solar PV capacity (400 GW) by 2030 (unit: 10,000 kW)

No.	Region	2015	2020	2030
1	Receiving ends in eastern and central China	1265	5200	12,500
2	Northeast China	84	800	1600
3	Northwest China	2083	5000	11,500
4	Sichuan, Chongqing, and Xizang	53	800	4000
5	Shanxi	113	500	1800
6	Western Nei Mongol	412	1000	3100
7	Southern China	148	1700	5500
Total		4158	15,000	40,000

From the perspective of energy replacement, the reduction of nuclear power 30 GW represents non-fossil energy of about 210 billion kWh. Assuming that the annual wind power utilization time is 1800–2000 h and the annual solar power utilization time is 1100–1300 h, the installed capacity of wind power and solar power needs to increase by about 50 and 100 GW, respectively.

5.3.2.4 Scenario 4: Demand slowdown scenario

Representative scenario—low power demand plan: The development scale of hydropower, nuclear power, wind power, and solar power can meet the expectations.

Research on the low power demand plan. Under the circumstances of low demands, non-fossil energy will continue to develop at a high rate driven by the policies, and the energy sources affected by low demand are mainly conventional thermal power, including coal power and gas power.

According to the low power demand scenario (China's power consumption is 7.0 trillion kWh in 2020), by 2020, China's installed generation capacity in 2020 is shown further: hydropower (345 GW), nuclear power (58 GW), wind power (200 GW), solar power (100 GW), and biomass power (17 GW), and in 2030 is shown further: hydropower (430 GW), nuclear power (130 GW), wind power (400 GW), solar power (300 GW), and biomass power (30 GW). The development plan of power generation based on non-fossil energy sources under scenario 2 is the same as that in scenario 1.

5.4 ANALYSIS ON SCENARIO OPTIMIZATION

5.4.1 SCENARIO 1: EXPECTED DEVELOPMENT SCENARIO

5.4.1.1 Installed power capacity

In 2015, China's installed power capacity reached 1.51 billion kW. Through analyses based on the high power demand plan, it is expected to reach 1.99 billion kW by 2020 and about 2.75 billion kW by 2030, showing a continuous and rapid rising tendency, see Table 5.13.

Table 5.13 The total power capacity under scenario 1 (unit: 10,000 kW)

Type of power source	2015	2020	2030
Coal	91,094	112,400	116,000
Fuel gas	6637	11,152	20,063
Hydro	29,666	34,523	43,160
Pumped storage	2271	3581	10,000
Nuclear	2717	5824	13,055
Wind	12,830	20,000	40,000
Solar	4158	10,000	30,000
Biomass	1300	1700	3000
Total	150,673	199,180	275,278

As is seen from the power source structure, the proportion of China's installed coal power capacity will keep declining in the future, down by 4 percentage points between 2016 and 2020, and 14.2 percentage points between 2020 and 2030. The proportion of installed fuel gas power capacity will rise steadily, and that of installed hydropower capacity will decline. Particularly, the proportions of the installation capacity based on new energy sources, including wind power and solar power, will rise most rapidly. By 2015, 2020, and 2030, the proportion of installed clean energy capacity will be 34%, 36%, and 47%, respectively. The power source structures in 2015, 2020, and 2030 are shown in Fig. 5.3, Fig. 5.4, and Fig. 5.5, respectively.

By type, the installed capacity and distribution of coal power, pumped storage and fuel gas power are shown further.

5.4.1.1.1 Coal power

China's installed coal power capacity was 910 GW in 2015. It is expected to reach 1120 GW and 1160 GW in 2020 and 2030, respectively. The plans for and proportions of conventional installed coal power capacity in the power grid areas are shown in Table 5.14 and Fig. 5.6, respectively.

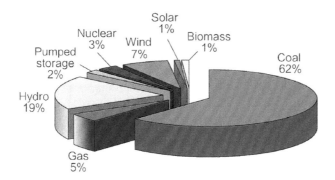

FIG. 5.3

The power source structure under scenario 1 (2015).

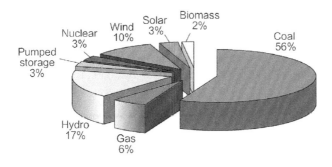

FIG. 5.4

The power source structure under scenario 1 (2020).

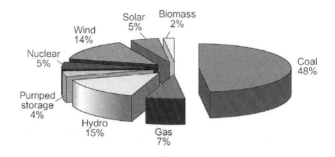

FIG. 5.5

The power source structure under scenario 1 (2030).

Table 5.14 The plan on conventional installed coal power capacity (unit: 10,000 kW)

No.	Region	2015	2020	2030
1	Receiving ends in eastern and central China	45,934	52,869	47,859
2	Northeast China	8414	9427	9800
3	Northwest China	10,752	17,600	22,571
4	Sichuan, Chongqing, and Xizang	2676	3500	3880
5	Shanxi	5645	7175	9355
6	Western Nei Mongol	5719	7862	8962
7	Southern China	11,954	13,967	13,573
Total		91,094	112,400	116,000

Between 2016 and 2030, the proportion of the installed coal power capacity of Shanxi, western Nei Mongol, and northwest China to the total installed coal power capacity of China will rise steadily, the proportion of the installed coal power capacity in the receiving ends in eastern and central China, Sichuan, Chongqing, Xizang, and southern China will drop gradually, and the proportion of the installed coal power capacity in northeast China to the total installed coal power capacity of China will remain at about 8.4%.

5.4.1.1.2 Pumped storage

China's installed pumped storage capacity stood at 22.71 GW in 2015 and is expected to reach 36 GW by 2020 and 100 GW by 2030. The plans for and proportions of pumped storage power station in the power grid areas are shown in Table 5.15 and Fig. 5.7, respectively.

At present, pumped storage power stations are mainly located at the power receiving ends in eastern and central China and southern China; during the "13th Five-year Plan" period, the construction of pumped storage power stations at the power receiving ends in eastern and central China and northeast China will accelerate; after 2020, the installed capacity of pumped storage power stations will take up a certain portion in the total installed capacity of the power grid areas.

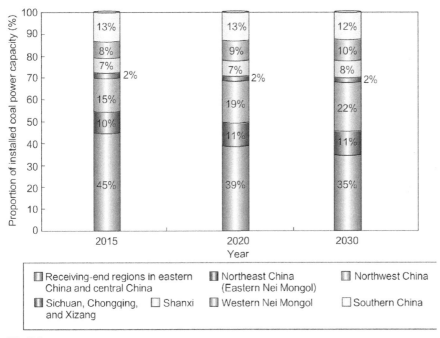

FIG. 5.6

A schematic diagram of proportions of installed coal power capacity in different power grid areas.

Table 5.15 The plan on installed capacity of pumped storage power stations (unit: 10,000 kW)

No.	Region	2015	2020	2030
1	Receiving ends in eastern and central China	1392	2142	5232
2	Northeast China	150	350	980
3	Northwest China	0	0	840
4	Sichuan, Chongqing, and Xizang	9	9	240
5	Shanxi	120	120	240
6	Western Nei Mongol	120	120	360
7	Southern China	480	840	2108
Total		2271	3581	10,000

5.4.1.1.3 Gas power station

China's installed capacity of fuel gas power was 66.37 GW in 2015. It is expected to reach 110 GW in 2020 and 200 GW in 2030. The plans for and proportions of installed gas power capacity in different grid areas are shown in Table 5.16 and Fig. 5.8, respectively.

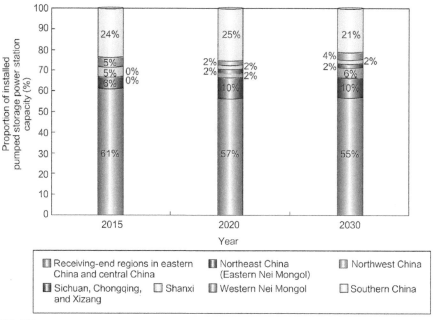

FIG. 5.7

A schematic diagram of proportions of installed capacity of pumped storage power stations in different power grid areas.

Table 5.16 The plan on installed gas power capacity (unit: 10,000 kW)

No.	Region	2015	2020	2030
1	Receiving ends in eastern and central China	4706	6654	12,564
2	Northeast China	34	210	510
3	Northwest China	121	430	704
4	Sichuan, Chongqing, and Xizang	255	294	519
5	Shanxi	268	452	522
6	Western Nei Mongol	0	126	308
7	Southern China	1253	2986	4936
Total		6637	11,152	20,063

Between 2016 and 2030, the installed capacity of gas power stations at the receiving ends in eastern and central China will take up the largest portion of the total installed capacity of gas power stations; the northeast China and northwest China power grids will show a rising tendency; the installed fuel gas capacity in Shanxi, Sichuan, Chongqing, and Xizang will also grow to a certain extent.

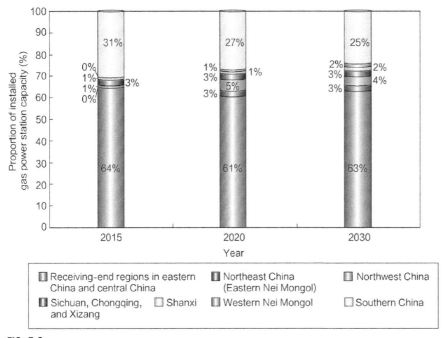

FIG. 5.8

A schematic diagram of proportions of installed gas power capacity in different power grid areas.

5.4.1.2 Development and consumption of clean energy

5.4.1.2.1 Cross-regional power flow

In the future, China will continue to show an overall tendency of power flow "from the west to the east and from the north to the south," and the power flow scale will rise markedly. In 2015, eastern and central China received a power flow of about 110 GW, including coal power of 65 GW, hydropower of 43 GW, and wind power of 5 GW.

In 2020, the power flow to eastern and central China is expected to rise to 220 GW, including coal power of 135 GW, hydropower of 65 GW, wind power of 54 GW, and solar power of 21 GW. The main power flows are shown in Table 5.17.

In 2030, the power flow to eastern and central China is expected to rise to 330 GW, including coal power of 180 GW, hydropower of 100 GW, wind power of 130 GW, and solar power of 120 GW. The main power flows are shown in Table 5.18.

Cross-provincial and cross-regional power transmission may help to expand the scope of clean energy development and consumption, give full play to the advantages of the large market and strong capability of power source regulation at the power-receiving ends in eastern and central China, increase the clean energy development scale, and enhance the utilization efficiency.

Table 5.17 The scale and direction of power flow in 2020 (unit: 100 GW)

No.	Region	Coal power	Hydropower	Wind power
1	Receiving ends in eastern and central China	1.35	0. 65	0.54
2	Northeast China	−0.1	–	−0.12
3	Northwest China	−0.45	–	−0.3
4	Sichuan, Chongqing, and Xizang	0.1	−0.68[a]	–
5	Shanxi	−0.3	–	–
6	Western Nei Mongol	−0.6	–	−0.12
7	Southern China	–	0.03	–

[a]Including the electricity of 22.4 GW generated from the Three Gorges Hydropower Station.

Table 5.18 The scale and direction of power flow in 2030 (unit: 100 GW)

No.	Region	Coal power	Hydropower	Wind power
1	Receiving ends in eastern and central China	1.8	1	1.3
2	Northeast China	−0.24	–	−0.32
3	Northwest China	−0.67	–	−0.68
4	Sichuan, Chongqing, and Xizang	0.1	−1.19[a]	–
5	Shanxi	−0.4	–	–
6	Western Nei Mongol	−0.6	–	−0.3
7	Southern China	–	0.19	–

[a]Including the electricity of 22.4 GW generated from the Three Gorges Hydropower Station.

5.4.1.2.2 Clean energy consumption and utilization

As is seen from wind power consumption, on the basis of the scale and distribution of China's wind power development, China's hydropower stations will concentrate mainly in the "Three Norths" regions, where the wind power consumption capability is limited and cannot meet the needs for large-scale wind power development. Therefore, it is necessary to speed up the construction of cross-provincial, cross-regional power grids so as to increase the scope and scale of wind power consumption. Research results show that, assuming that China's installed wind power capacity is 200 GW in 2020, 96 GW will be consumed within the province, 14 GW will be consumed in other places in the region, and the remaining 90 GW will be consumed beyond the region. As is seen from the way of wind power transmission, the coal power generated from northwest China and western Nei Mongol will be delivered out of the region on a large scale, while the wind power generated from northwest China and western Nei Mongol may be delivered out of the region, together with coal power, via the corridors for coal power transmission. In northeast China, wind power will grow on a large scale, according to the need for large-scale cross-regional transmission. However, the power transmission corridors in northeast China are mainly used for wind power transmission, and thus have great operational difficulty and economic costs compared with the joint wind-thermal power transmission in northwest China and western Nei Mongol. As is seen from wind power utilization, the expanding cross-regional power flow will help realize cross-regional wind power transmission, and the wind curtailment ratio will be 4.8%. The major wind power consumption market and major wind power-producing provinces by 2020 are shown in Table 5.19, and the proportions of wind power consumption in the power grid areas are shown in Fig. 5.9.

Around 2030, China will lay equal stress on the development of onshore wind power and offshore wind power and carry out demonstration wind power projects on the high seas. Meanwhile, it will speed up development of distributed wind power in its eastern and central regions, and build large wind power bases in the coastal areas in Shandong, Shanghai, Zhejiang, Guangdong, and Fujian; the wind power generated from eastern and central China and southern China is largely consumed locally, and the wind power flow from western and northern China will grow steadily. By 2030, China's wind power transmission for nonlocal consumption will reach 130 GW, and the wind curtailment of China will reach about 3.5%, a relatively low level.

As is seen from solar energy consumption and utilization, before 2020, China's installed solar power capacity is relatively small. In northwest China, which has the largest installed solar power capacity of the country, the proportion of solar generation to China's total power consumption is about 4.5%, a relatively low level; in the case of the power flow under scenario 1, the proportion of solar curtailment will be about 2.5%, mainly occurring in northwest China.

Table 5.19 Major wind power-producing provinces and wind power consumption market by 2020 (unit: 10,000 kW)

Region	Development scale	Local consumption	Cross-provincial power transmission	Cross-regional transmission
Receiving ends in eastern and central China	550	2900	26	
Northeast China	340	1600	4	1600
Northwest China	540	1100	8	3500
Sichuan, Chongqing and Xizang	70	700	0	0
Shanxi	80	800	0	0
Western Nei Mongol	2400	700	2	1500
Southern China	1800	1800	0	0
Total	20,000	9600	14	9000

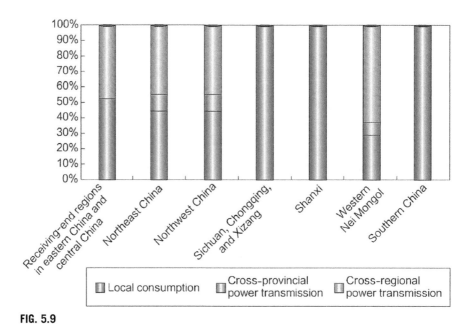

FIG. 5.9

A schematic diagram of proportions of wind power consumption in different power grid areas.

In 2030, China will have the preliminary capabilities of manufacturing of key equipment, as well as large-scale design, construction, and maintenance. In terms of the distribution of solar power stations, distributed PV will pick up pace in the eastern and central regions of China. Moreover, China will build 1 GW rooftop PV systems in Beijing, Tianjin, Shanghai, Jiangsu, Guangdong, Yunnan, and Shandong. The cross-regional power transmission demand of the solar generation bases in northwest China will steadily rise. Estimates show that the installed capacity of solar power that needs to be delivered for consumption will reach 120 billion kW by 2030, and the proportion of solar curtailment will be about 0.5%.

5.4.1.3 Analysis on the typical operation mode of the system

Northwest China has abundant hydroenergy, wind energy, and solar energy resources. Among the seven large power grid areas, northwest China ranks second in terms of installed capacity of hydropower and wind power and ranks first in terms of installed solar power capacity. Northwest China has a great variety of non-fossil energy sources, which are typical and on a large scale. Therefore, we use northwest China as an example to analyze the typical ways of system operation.

5.4.1.3.1 Overall system operation

Considering the cross-regional power transmission in northwest China, in accordance with the requirements for giving priority to renewable energy dispatching and ensuring the economical dispatching and operation of the electric power system, a simulation analysis is made on system production. On the whole, both the minimum technical output of thermal power and the forced output of hydropower operate at the position of base load, and wind power and PV power generation shall be given priority in the remaining power generation space of the system. During the period of load valley, in the conditions that both hydropower and thermal power are at the level of maximum output and the pumped storage is under the status of water pumping, due to the sufficient supply of wind power, the method for wind curtailment must be used during certain periods so as to ensure system balance. The hourly output curve of various units of power grids in northwest China on a typical day in 2030 is shown in Fig. 5.10.

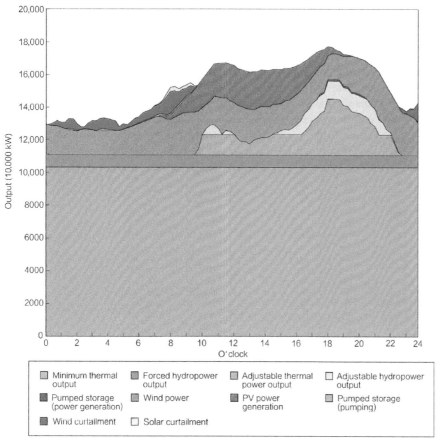

FIG. 5.10

Simulation of system operation of the Northwest China power grid on a typical winter day in 2030 (including power transmission).

5.4.1.3.2 Operation and utilization of wind power

By considering the regulating capability of various units during different periods on the basis of the hourly operating positions of various power sources, the wind power that can be balanced by the system during different periods can be analyzed. Generally speaking, the system has a higher capability of balancing the wind power in the daytime and tends not to give rise to wind curtailment; the system has a lower capability of balancing the wind power at night, and if the installed wind power capacity exceeds the wind power receiving capability of the system, it may give rise to wind curtailment, as is shown in Fig. 5.11.

By 2020, the installed wind power capacity of northwest China is expected to reach 54GW. If the cross-regional power grid construction is considered, the whole-year wind curtailment ratio of the northwest China power grid will reach about 6.2% in the year. Specifically, due to heat supply in winter, the thermal power regulation capability is limited. Moreover, the proportion of wind curtailment is high due to the lower system load level as well as such factors as the overhauling of hydropower stations, etc. Particularly, in the windy months of January and February with the lowest load, and the proportion of wind curtailment will exceed 10%. As is shown in Fig. 5.12, in northwest China, heavy rainfall in summer and sufficient runoff hydropower generation that cannot be regulated and will also cause certain wind curtailments.

By 2030, the installed wind power capacity of northwest China is expected to rise to 100GW, and the cross-provincial cross-regional wind power consumption will further increase with the construction of cross-regional power transmission corridors. As is shown in Fig. 5.13, the whole-year wind curtailment ratio in northwest

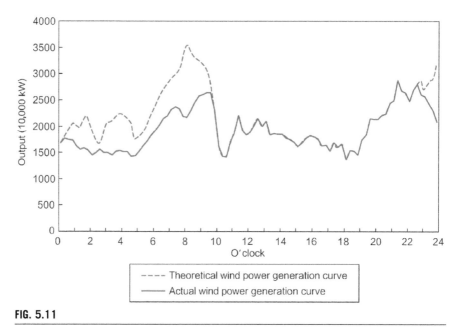

FIG. 5.11

The typical operation mode of a wind power station.

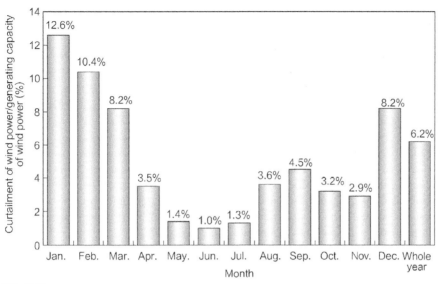

FIG. 5.12

A schematic diagram of wind power consumption and wind curtailment of power grids in northwest China in 2020 (including power transmission).

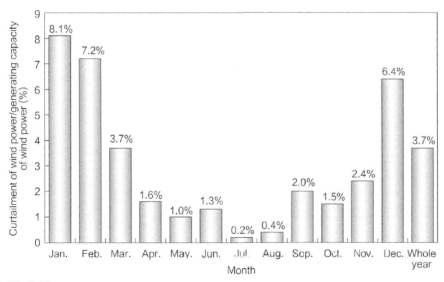

FIG. 5.13

A schematic diagram of wind power consumption and wind curtailment of power grids in northwest China by 2020 (including power transmission).

China will be around 3.7% by 2030, and there is still a relatively high-wind curtailment ratio during a winter period with sufficient supply of wind power.

5.4.1.3.3 Operation and utilization of solar power

As a result of the rapid growth of the installed solar PV capacity in northwest China, the constraints of the system regulation capability on solar generation will deliver results. Under the power source structure of the expected scenario and with the typical winter operation mode in northwest China, the load level in northwest China is still relatively low at 07:00–10:00, the system regulating capability is relatively small, the wind power output still remains relatively large, and the solar power output has entered a stage of gradual increase. Under such circumstances, the inadequate system adjustment capability may also cause wind curtailment and solar curtailment simultaneously, as is shown in Fig. 5.14. In general, since installed solar PV capacity is relatively small, solar curtailment is not prominent all year round. The solar curtailment ratio will be about 2.5% in 2020 and about 0.5% in 2030.

5.4.1.3.4 Operation mode of pumped storage

Pumped storage has the function of peak load shaving in the electric power system. If the wind power and solar power generation in the system are on a small scale, pumped storage stations shall pump water at a period of low load (about 05:00) and generate electricity at a period of peak load (about 20:00). However, the large-scale development of wind power and solar power has changed the low-load

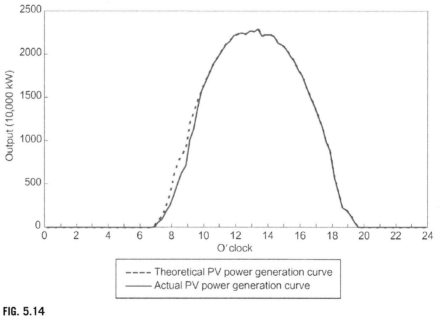

FIG. 5.14

A schematic diagram of a typical operation mode of solar power stations.

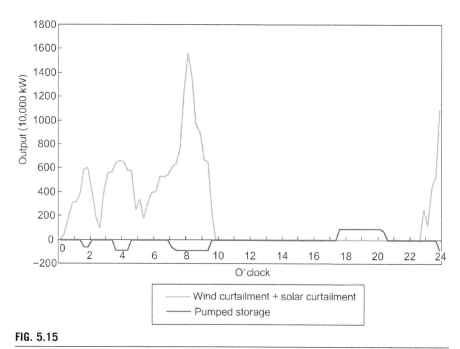

FIG. 5.15

The typical operating mode of pumped storage.

period of the system. Pumped storage power stations, on the basis of need for real-time system balance, pumps water at the most difficult time of system peak shaving (the largest wind/solar curtailment), so as to give play to the function of valley filling and use renewable power to the greatest extent. As is seen from Fig. 5.15, after wind power and PV power are connected to the grid, the low-valley period and the peak shaving demand period of the system will change. At 01:00, 04:00, and the period between 07:00 and 10:00, the wind power and PV output are relatively large, with the most difficult system peak shaving, and the pumped storage power stations will pump in those periods.

5.4.1.3.5 The promoting effects of cross-regional transmission on clean energy consumption and utilization

Without the large-scale cross-regional power transmission, the equivalent load level of northwest China will drop, and the system operation scale and the regulation capability will decline. If the installed capacity of wind power and solar power remains unchanged, the proportion of system wind/solar curtailment will increase significantly. If the cross-regional transmission is not considered, an hourly analysis on the output curves and operating positions of various units is made on the same typical day in accordance with the requirements of giving priority to renewable energy dispatching, economical dispatching, and the operation of the electric power system. The analysis is shown in Fig. 5.16.

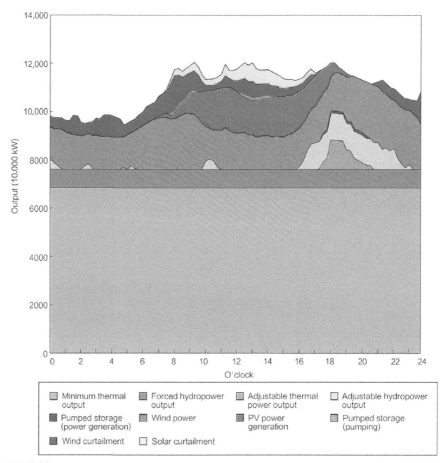

FIG. 5.16

System operation simulation of the power grid in northwest China on a typical winter day in 2030 (excluding power transmission).

By contrasting the system operation and simulation situations of the power grid in northwest China on a typical winter day (the situation with outward power transmission against that without outward power transmission), we can see that, if power transmission is not considered, the time and electricity of wind curtailment in the power grid in northwest China will increase considerably. Analysis results show that, if power transmission is not considered, the whole-year wind curtailment proportion of the power grid in northwest China will reach 14.8%, and the proportion of wind power curtailment, during some months of difficult power consumption, may approach 40%. Compared with the situation with cross-regional transmission, the wind power curtailment will increase by 12 billion kWh. The wind power

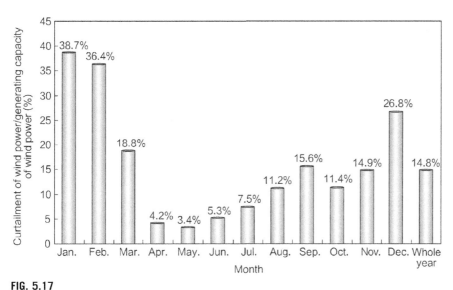

Wind power consumption and wind curtailment of the power grid in northwest China in 2030 (no power transmission).

consumption and wind curtailments of the power grid in northwest China by 2030 are shown in Fig. 5.17.

As is seen from the consumption and utilization of solar energy, by contrasting the system operation and simulation situations of power grid in northwest China on a typical winter day (the situation with outward power transmission against that without outward power transmission), we can see that, if cross-regional transmission is not considered, the proportion of solar curtailment will increase. The research results show that, if cross-regional transmission is not considered, the proportion of solar curtailment in northwest China will reach 5.7%, and the solar curtailment will increase by 2 billion kWh compared with the situation for which cross-regional transmission is considered.

5.4.1.4 Analysis on economical efficiency

The total cost and composition of China's electric power supply between 2011 and 2030 under the expected development scenario are shown in Table 5.20.

As is seen from the composition of the total costs, within the 20-year calculation period, in the total costs, investments account for about 29%, fuel costs account for about 57%, fixed operating expenses account for about 10%, and external environmental expenses account for about 4%, as is shown in Fig. 5.18. On the one hand, the large-scale development of wind power and PV power generation will give rise to the investment scale, on the other hand, the electricity generated will save the operating fuel costs of the system and reduce the emissions of environmental pollutants. The development of new energy sources will replace coal and lower external

Table 5.20 Analysis on the economical efficiency of the expected development scenario (scenario 1) (unit: RMB 100 million yuan)

Economic indicators	Value
Total cost	268,547
1. Investment	78,177
1.1 Wind	12,832
1.2 Nuclear	9317
1.3 Gas	3052
1.4 Pumped storage	1550
1.5 Hydro	9810
1.6 Coal	9554
1.7 Solar and biomass[a]	17,417
1.8 Power grid	14,645
2. Fuel cost	152,933
3. Fixed operating expenses	27,325
4. External environmental expenses	10,112

Note: *Based on the calculation period of 2011–30, with a discount rate of 8%, and the investments in the construction of power transmission network and power distribution network of the provinces are not covered in the power grid investment.*
[a]*The investment in other power sources includes the investment in the installed capacity of solar power generation and biomass power generation.*

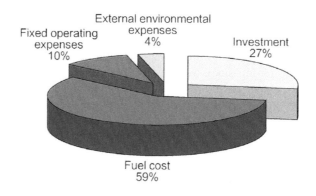

FIG. 5.18

The composition of total system costs between 2011 and 2030.

environmental expenses, and by its very nature, is achieving the benefits of coal power by making investments in power sources, power grids, and operations, as a result of the development of new energy power generation. As is seen from the current price level, the investment in new energy sources including wind power is relatively high, and its benefits can only offset a part of its initial investment, and its economical efficiency needs to be further improved.

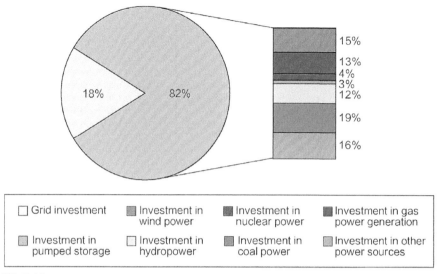

FIG. 5.19

The composition of power investments between 2011 and 2030.

As is seen from this type of investment, the proportion of the investments in power sources and power grids is about 4:1%, as is shown Fig. 5.19. If all investments in power grid construction are considered, the proportion is about 4:3%. In power source investments, the proportion of clean energy investment will reach about 78%. Therefore, the Chinese government needs to give priority to investment in the installed capacity of clean energy sources such as hydropower, nuclear, wind, solar, and biomass energy sources.

5.4.2 SCENARIO 2: HYDROPOWER SLOWDOWN SCENARIO

5.4.2.1 Installed power capacity

Under the hydropower slowdown scenario, the reduced hydropower needs to be made up by wind power and solar power. Since the capacity substitution effect of the installed capacity for power generation, based on new energy sources, including wind power and solar PV power, is very small, the reduced hydropower capacity also needs the increase of power sources such as coal power, fuel gas, and pumped storage with the equivalent capacity substitution effects. Therefore, the reduction of installed hydropower capacity will give rise to the increase of the total installed capacity of the system power source. By 2020, the overall installed hydropower capacity is expected to reach nearly 2.03 billion kW, about 40 GW higher than that under scenario 1; by 2030, the overall installed hydropower capacity is expected to reach about 2.85 billion kW, about 98 GW higher than that under scenario 1. The specific installed power capacity is shown in Table 5.21.

Table 5.21 The total power capacity under scenario 2 (unit: 10,000 kW)

Type of power source	2015	2020	2030
Coal	91,094	113,000	116,300
Fuel gas	6637	11,652	21,489
Hydro	29,666	33,223	39,220
Pumped storage	2271	3581	12,020
Nuclear	2717	5824	13,055
Wind	12,830	22,000	45,000
Solar	4158	12,000	35,000
Biomass	1300	1700	3000
Total	150,673	202,980	285,084

As is seen from the power source structure, as a result of hydropower development slowdown, the proportion of installed hydropower capacity to the total installed capacity will further decline, and the installed pumped storage and fuel gas power capacity as a substitute will rise to a certain extent. The installed power capacity in 2020 and 2030 are shown in Fig. 5.20 and Fig. 5.21, respectively.

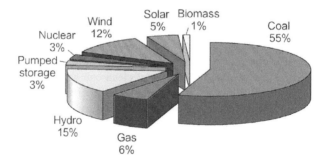

FIG. 5.20

The power source structure under scenario 2 (2020).

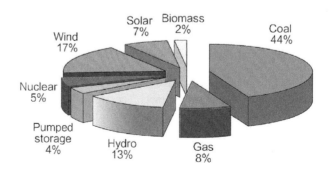

FIG. 5.21

The power source structure under scenario 2 (2030).

Table 5.22 The plan on conventional installed coal power capacity (unit: 10,000 kW)

No.	Region	2015	2020	2030
1	Receiving ends in eastern and central China	45,934	53,269	48,159
2	Northeast China	8414	9427	9800
3	Northwest China	10,752	17,600	22,571
4	Sichuan, Chongqing, and Xizang	2676	3600	3880
5	Shanxi	5645	7175	9355
6	Western Nei Mongol	5719	7862	8962
7	Southern China	11,954	14,067	13,573
Total		91,094	113,000	116,300

By type, the installed capacity of coal power, gas power, and pumped storage will rise. Specifically, the installed coal power capacity will increase by 6 GW by 2020, mainly in the receiving-end provinces. By 2030, the installed capacity of hydropower will mainly be replaced by that of pumped storage and fuel gas, and the installed coal power capacity will remain unchanged. The coal power development in power grid areas is shown in Table 5.22.

By 2020, the hydropower development slowdown will cause an increase of installed wind power capacity and installed solar power capacity by 20 GW, respectively. On the one hand, the new renewable energy consumption needs the construction of synchronous receiving-end power grids, reduction of the system peak-valley differences, and the enhancement of the renewable energy receiving capability; on the other hand, the power transmission curve between the sending ends and the receiving ends have a direct impact on the distribution of peak shaving power sources and the change of the overall power source structure. In order to increase the scale of power transmission from wind power bases, it is necessary to appropriately reduce the involvement of power transmission in peak shaving of the receiving-end power grid. Therefore, it is necessary to increase the construction scale of peak shaving power sources at the receiving-end power grids.

By considering the above factors, through system optimization analyses, if the wind power development scale reaches 450 GW by 2030, compared with the wind power development scale of 400 GW under scenario 1, it is necessary to increase the installed pumped storage capacity by about 20 GW and the installed gas power capacity by about 14 GW by 2030. The plans on the installed capacity of pumped storage power stations and gas power stations are shown in Tables 5.23 and 5.24, respectively.

5.4.2.2 Clean energy consumption
5.4.2.2.1 Overall changes of cross-regional power flow
Compared with scenario 1, under scenario 2, by 2030, the hydropower development scale of Sichuan, Chongqing, and Xizang will drop by 37 GW, and that of northwest China will drop by 5 GW, making an impact of the hydropower power flow in the

Table 5.23 The plan on installed capacity of pumped storage power stations (unit: 10,000 kW)

No.	Region	2015	2020	2030
1	Receiving ends in eastern and central China	1392	2142	5972
2	Northeast China	150	350	1180
3	Northwest China	0	0	100
4	Sichuan, Chongqing, and Xizang	9	9	560
5	Shanxi	120	120	360
6	Western Nei Mongol	120	120	480
7	Southern China	480	840	2468
Total		2271	3581	12,020

Table 5.24 The plan on installed gas power capacity (unit: 10,000 kW)

No.	Region	2015	2020	2030
1	Receiving ends in eastern and central China	4706	6854	13,342
2	Northeast China	34	260	510
3	Northwest China	121	480	804
4	Sichuan, Chongqing, and Xizang	255	294	619
5	Shanxi	268	452	622
6	Western Nei Mongol	0	126	308
7	Southern China	1253	3186	5284
Total		6637	11,652	21,489

next 10 years by about 30 GW, while the coal power flow will remain basically unchanged. Meanwhile, due to the significant increase of wind power and solar power development scale and the limited local consumption, larger-scale cross-regional power transmission will be needed. Under this scenario, the scale of wind power and solar power flow will rise remarkably, and the wind power flow demand for transmission will increase by about 30 GW by 2030, as is shown in Table 5.25. The total power flow to the receiving ends in eastern and central China will reach 350 GW by 2030.

5.4.2.2.2 Clean energy consumption and utilization

As is seen from wind power consumption, under scenario 2, cross-regional transmission corridors need to deliver more electricity generated from new energy, and the coal power transmission from western China and northern China will fall compared with that under scenario 1. The electricity generated, based on new energy, is increased mainly by changing the operation methods of power transmission lines and the scale of supporting thermal power, wind power, and solar power generation. Since the wind power generation in "Three Norths" regions and the solar power generation in western China have increased significantly, the proportion of wind and

Table 5.25 The power transmission demand under the hydropower slowdown scenario by 2030 (unit: 100 GW)

No.	Region	Coal power	Hydropower	Wind power
1	Receiving ends in eastern and central China	1.8	0.8	1.6
2	Northeast China	−0.24	–	−0. 32
3	Northwest China	−0.67	–	−0. 88
4	Sichuan, Chongqing, and Xizang	0.1	−0.9	–
5	Shanxi	−0.4	–	–
6	Western Nei Mongol	−0.6	–	−0.4
7	Southern China	–	0.1	–

solar curtailments will rise markedly. The average proportion of wind curtailment of China will reach about 9.6%. Specifically, the proportion of wind curtailment in "Three Norths" will exceed 10%, up about 6 percentage points compared with that under scenario 1.

As is seen from solar energy consumption and utilization, after the reduction of the hydropower development scale, as a result of large-scale increase of solar generation and the possible impacts of large-scale wind power development on solar power consumption, the proportion of solar curtailment will reach about 3.2% by 2030, up by 2.7 percentage points compared with that under scenario 1.

5.4.2.3 Analysis on typical ways of system operation
5.4.2.3.1 Overall system operation
Considering the cross-regional power transmission in northwest China, in accordance with the requirements on giving priority to renewable energy dispatching and ensuring economical dispatching and operation of the electric power system, a simulation analysis is made on system production. On the whole, due to the increase of the wind and solar power generation scale, the wind and solar curtailments will also increase. In addition, compared with scenario 1, due to the large-scale increase of solar generation, solar power output is higher in the daytime, and the pumped storage power stations may pump water in the daytime. The curve of hourly output of various units of the power grid in northwest China on a typical day is shown in Fig. 5.22.

5.4.2.3.2 Operation and utilization of wind power
As is seen from wind power consumption and utilization, compared with scenario 1, the installed wind power capacity of northwest China will increase to 110 GW in 2030, up by 100 GW. As is seen from the influence of the change factors, the reduced hydropower scale will reduce the regulating capability of the power gird in northwest China, and the increased PV power generation scale will have an impact on the wind power consumption between 07:00 and 10:00 each day. Although the installed

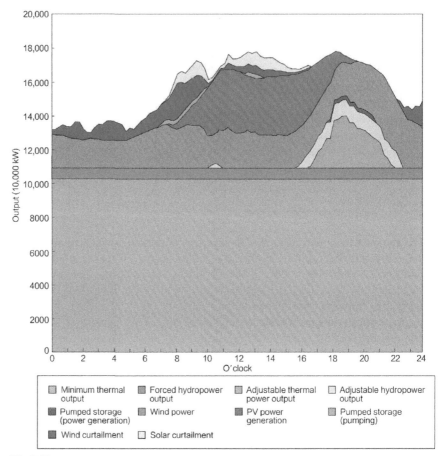

FIG. 5.22

Simulation of the system operation of the power grid of in northwest China on a typical winter day in 2030 (including power transmission).

capacity of pumped storage power stations will rise to 10 GW, it cannot meet the needs of new wind power and PV regulation, and the proportion of overall wind curtailment will continue to rise in northwest China. Research results indicate that, under the power source structure of scenario 2, the proportion of wind curtailment in northwest China will rise to 11.8%, and is even likely to approach 30% in certain months in winter, as is seen from Fig. 5.23.

5.4.2.3.3 Operation and utilization of solar power

As is seen from PV consumption and utilization, compared with scenario 1, by 2030, the installed PV power capacity in northwest China will increase from 100 to 107 GW, with a peak PV output of around 40 GW, accounting for over 30% of the maximum system load, higher than the peak-valley difference (about 20%) of

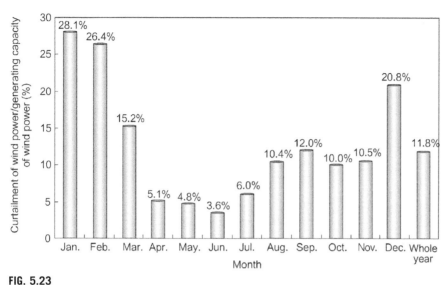

FIG. 5.23

A schematic diagram of wind power consumption and curtailment of the power grid in northwest China by 2030.

the load curve in northwest China. The PV power generation reaches its peak at noon each day, making a fundamental impact on the load curve in northwest China. The valley period of the load curve for northwest China is generally at night, since solar generation takes up a very high portion of the entire energy consumption. The equivalent load curve of northwest China has two valleys at noon and midnight, respectively, as is shown in Fig. 5.24. Therefore, in the case of a high proportion of wind power and solar power, the system peak shaving is difficult, and it is inevitable for wind curtailment to occur after midnight and solar curtailment to occur at noon. Analysis results show that, in the power source structure under scenario 2, the proportion of solar curtailment in northwest China will be about 5.2% in 2030.

5.4.2.4 Comparison of economical efficiency

A comparison of the economical efficiency between scenario 1 (expected development scenario, the goals for non-fossil energy sources will have been met by 2030) and scenario 2 (the hydropower development slows down by 2030) has been made, see Table 5.26 for detailed data of comparison of the economical efficiency.

The conclusions have been made in accordance with the calculation results of system optimization.

Under scenario 2, the net increase of the investment in the electric power system is RMB 741.4 billion. Specifically, the investment in wind power and solar power increases by RMB 806 billion, due to the increase of the development scale; the investment in pumped storage and fuel gas units increase by RMB 63.5 billion due to the increased installed capacity; the investment in installed hydropower

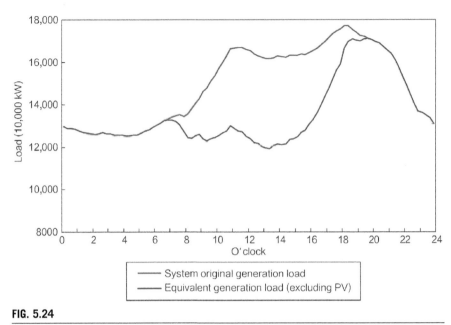

FIG. 5.24

The impacts of PV power generation grid connection on system load characteristics.

Table 5.26 Economical efficiency comparison under the hydropower slowdown scenario (unit: RMB 100 million yuan)

Economic indicators	Expected development scenario (scenario 1)	Hydropower slowdown scenario (scenario 2)	Change between scenario 1 and scenario 2
Total cost	310,673	320,667	9994
1. Investment	82,343	89,757	7414
1.1 Wind	12,832	17,172	4340
1.2 Nuclear	10,923	10,923	0
1.3 Gas	3052	3253	201
1.4 Pumped storage	2214	2648	434
1.5 Hydro	9810	7899	−1911
1.6 Coal	15,547	15,679	132
1.7 Other power sources	13,320	17,040	3720
1.8 Power grid	14,645	15,143	498
2. Fuel cost	184,568	185,624	1056
3. Fixed operating expenses	31,558	32,723	1165
4. External environmental expenses	12,204	12,563	359

Note: *based on the calculation period of 2011–30, with a discount rate of 8%, and the investments in the construction of power transmission networks and power distribution networks of the provinces are not covered in the power grid investment.*

capacity declines by RMB 191.1 billion due to the reduction of the installed hydropower capacity; under the two development scenarios, the installed capacity of nuclear power and that of biomass power remain basically unchanged, and the investment in nuclear power and in biomass power are basically the same; under scenario 2, the development scale of wind power and solar power increases, the cost of grid connection and power transmission increases, reduced hydropower development lowers the costs of grid connection and power transmission, and the net increase of power grid investment is about RMB 50 billion.

Under scenario 2, on the one hand, the increase of the development scale of wind power and solar power replaces more thermal power generation and reduces the consumption of power coal; on the other hand, the reduced hydropower development scale raises the fuel consumption. On the whole, due to the increase of wind curtailment and solar curtailment under this scenario and the reduced utilization efficiency of the installed capacity of non-fossil energy, the cost of power-generating fuel cost will increase by RMB 105.6 billion.

Under scenario 2, the fixed operating expenses of the electric power system will increase by RMB 116.5 billion due to the increase of installed power capacity and the scale of power grid construction.

Under scenario 2, the total electricity generated from non-fossil energy will drop, fuel consumption will rise, the emission of environmental pollutants will increase, and external environmental expenses will increase about RMB 35.9 billion.

On the whole, under scenario 2, the investment in power generation based on renewable energy (wind power and solar power) and the peak shaving power source (pumped storage, fuel gas) will rise, hydropower investment will decline, power grid investment will rise, the cost of power-generating fuel will go up, and fixed operating expenses and external environmental expenses will rise. In general, scenario 2 will cause China's total cost of electric power supply to rise by about RMB 1 trillion, as is shown in Fig. 5.25.

5.4.3 SCENARIO 3: NUCLEAR POWER SLOWDOWN SCENARIO

5.4.3.1 Installed power capacity

The development scale of nuclear power declines but that of wind power and solar power increases, so it is necessary to increase the construction scale of power sources for peak shaving in the receiving-end power grids. Since the effective capacity of wind power is very small and can only replace limited power capacity, and as the development scale of nuclear power drops, it is necessary to build coal power, pumped storage, and gas power units with a considerable capacity, so as to ensure system power and energy balance and meet the needs of peak shaving balance. It is estimated, that to reduce the installed nuclear power capacity by 30 GW by 2030, it is necessary to increase the installed capacity of coal power by 4 GW, that of pumped storage power station by 13 GW, and that of gas power stations by 14 GW. The plans for installed capacity of pumped storage power stations and gas power stations are shown in Tables 5.27 and 5.28, respectively.

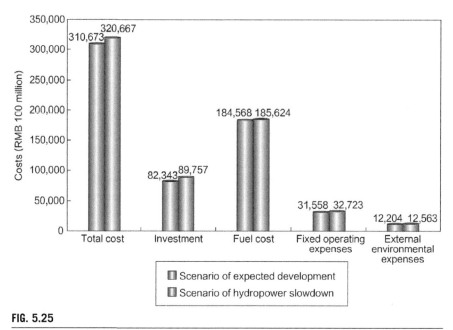

FIG. 5.25

The comparison of total costs between the hydropower slowdown scenario and the expected development scenario.

Table 5.27 The plan for installed capacity of pumped storage power station (unit: 10,000 kW)

No.	Region	2015	2020	2030
1	Receiving ends in eastern and central China	1392	2142	5872
2	Northeast China	150	350	980
3	Northwest China	0	0	960
4	Sichuan, Chongqing, and Xizang	9	9	240
5	Shanxi	120	120	240
6	Western Nei Mongol	120	120	480
7	Southern China	480	840	2528
Total		2271	3581	11,300

Table 5.28 The plan for installed gas power capacity (unit: 10,000 kW)

No.	Region	2015	2020	2030
1	Receiving ends in eastern and central China	4706	6970	13,322
2	Northeast China	34	310	510
3	Northwest China	121	530	704
4	Sichuan, Chongqing, and Xizang	255	500	519
5	Shanxi	268	400	522
6	Western Nei Mongol	0	126	308
7	Southern China	1253	3286	5528
Total		6637	12,122	21,413

5.4.3.2 Clean energy consumption
5.4.3.2.1 Overall changes of power flow
Under the nuclear power slowdown scenario, by 2030, the installed capacity of coal power, hydropower, and wind power will see little change, and the total power flow under this scenario is equivalent to that under scenario 1. In the long term, by 2030, as the development scale of nuclear power further drops, the installed capacity of solar power as an alternative energy will go up and bring about greater demands for cross-regional transmission, it is predicted that solar generation power flow will increase by 20–140 GW, compared with that under scenario 1.

5.4.3.2.2 Clean energy consumption and utilization
As is seen from wind power consumption, under scenario 3, the scale of coal power flow in western and northern China will decline compared with that under scenario 1, whereas the installed wind power capacity and the proportion of wind curtailment of China will slightly increase. On average, the proportion of wind curtailment of China is expected to reach about 4.1% by 2030. As is seen from the wind power consumption market, compared with that under scenario 1, China's cross-provincial, cross-regional wind power flow will increase by 15–145 GW.

As is seen from solar energy consumption and utilization, after the nuclear power development scale has been reduced, a certain solar curtailment may occur due to increased solar generation. It is predicted that the proportion of China's solar curtailment will be about 1.5% by 2030, concentrated mainly in northwest China.

5.4.3.3 Analysis on typical ways of system operation
5.4.3.3.1 Overall system operation
Considering the cross-regional power transmission in northwest China, in accordance with the requirements for giving priority to renewable energy dispatching and ensuring economical dispatching and operation of the electric power system, a simulation analysis is made on system production. The hourly output curves of various units on a typical day in 2030 are shown in Fig. 5.26.

5.4.3.3.2 Wind power operation and utilization
As is seen from wind power consumption and utilization, compared with scenario 1, the installed wind power capacity in northwest China will remain unchanged by 2030. As is seen from the change factors, the increase of installed solar PV capacity will also make a certain impact on the wind power consumption. Research results show that, in the case of the power source structure under scenario 3, the proportion of wind curtailment in northwest China will increase to 4.8%, and the proportion of wind curtailment in certain month(s) will reach about 15%, as is shown in Fig. 5.27.

5.4.3.3.3 Operation and utilization of solar power
As is seen from PV consumption and utilization, compared with scenario 1, the installed PV power capacity in northwest China will increase from 100 to 115 GW, and the proportion of solar curtailment will further rise. Analysis results

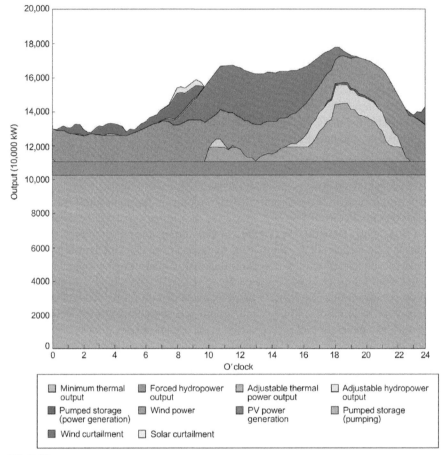

FIG. 5.26

System operation simulation of the power grid on a typical winter day in northwest China in 2030 (including power transmission).

show that, in the power source structure under scenario 3, the whole-year proportion of solar curtailment in northwest China will be 1.4% in 2030.

5.4.3.4 Comparison of economical efficiency

A comparison of the economical efficiency between scenario 1 (expected development scenario) and scenario 3 (nuclear power slowdown) has been made, as is shown in Table 5.29.

According to the above calculation results of system optimization, by estimate, under scenario 3, the increased development scale of renewable energy will increase the investments in solar generation, peak shaving power source, and gas power, lower the investment in nuclear power and raise the investment in power grids;

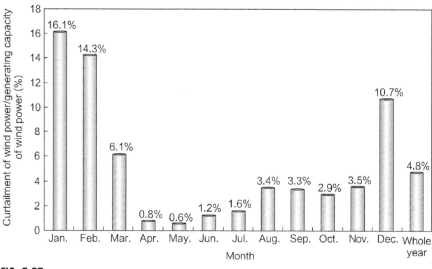

FIG. 5.27

A schematic diagram of wind power consumption and curtailment of the power grid of northwest China by 2030.

Table 5.29 Comparison of economical efficiency of nuclear power slowdown scenario (unit: RMB 100 million yuan)

Economic indicators	Expected development scenario (scenario 1)	Nuclear power slowdown scenario (scenario 3)	Change between scenario 1 and scenario 3
Total cost	310,673	314,502	3829
1. Investment	82,343	85,229	2886
1.1 Wind	12,832	13,002	170
1.2 Nuclear power investment	10,923	9603	−1320
1.3 Gas	3052	3269	217
1.4 Pumped storage	2214	2359	145
1.5 Hydro	9810	9810	0
1.6 Coal	15,547	15,848	301
1.7 Other power sources	13,320	16,524	3204
1.8 Power grid	14,645	14,814	169
2. Fuel cost	184,568	184,710	42
3. Fixed operating expenses	31,558	32,325	767
4. External environmental expenses	12,204	12,238	34

Note: based on the calculation period of 2011–30, with a discount rate of 8%, investments in the construction of power transmission networks and power distribution networks of the provinces are not covered in the power grid investment.

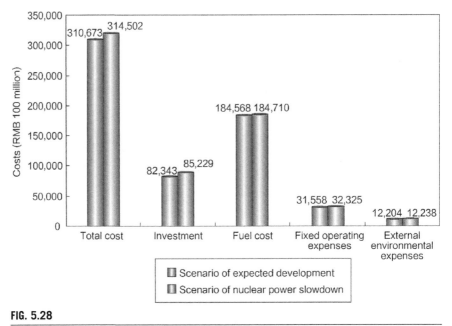

FIG. 5.28

Comparison of total costs under the nuclear power slowdown scenario and the expected development scenario.

the cost of power-generating fuel will remain basically unchanged, fixed operating expenses of system will increase, and external environmental expenses will increase to a certain extent. On the whole, compared with scenario 1, scenario 3 will increase the total cost of China's electric power supply by about RMB 383 billion, as is shown in Fig. 5.28.

5.4.4 SCENARIO 4: DEMAND SLOWDOWN SCENARIO

5.4.4.1 Calculation results of the plan on installed power capacity

Under scenario 4, due to the decline of power demands, China's total installed capacity will fall significantly to about 1860 GW by 2020, down by about 130 GW over that in scenario 1 by 2030, the total installed capacity will reach about 253 GW, down by about 220 GW compared with that under scenario 1. The plan for the installed power capacity under scenario 4 is shown in Table 5.30.

As is seen from the power source structure, since clean energy has reached the expected development scale, the proportion of installed clean energy capacity will increase from 36.2% to 38.8% in 2020, under scenario 1. It is predicted that the proportion of installed coal power capacity will further decline, down by 5.8 percentage points between 2016 and 2020, and by 15 percentage points between 2020 and 2030. The plans for installed power capacity in 2020 and 2030 under scenario 4 are shown in Figs. 5.29 and 5.30, respectively.

Table 5.30 The total power capacity under scenario 4 (unit: 10,000 kW)

Type of power source	2015	2020	2030
Coal	91,094	101,554	100,288
Fuel gas	6637	8528	13,022
Hydro	29,666	34,523	43,160
Pumped storage	2271	3581	10,290
Nuclear	2717	5824	13,055
Wind	12,830	20,000	40,000
Solar	4158	10,000	30,000
Biomass	1300	1700	3000
Total	15, 673	185,710	252,815

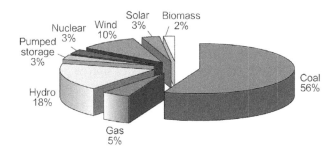

FIG. 5.29

The power source structure under scenario 4 (2020).

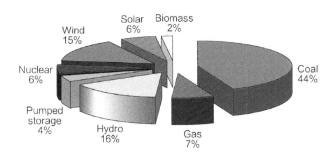

FIG. 5.30

The power source structure under scenario 4 (2030).

By type, the installed capacity and distribution of coal power, pumped storage, and fuel gas power are shown further.

5.4.4.1.1 Coal power

Under the scenario of low power demand, China's installed coal power capacity will reach 1.02 billion kW by 2020 and 1 billion kW by 2030, as is shown in Table 5.31. Compared with scenario 1, it will fall by 100 GW in 2020, and 160 GW in 2030.

Table 5.31 The plan on conventional installed coal power capacity (unit: 10,000 kW)

No.	Region	2015	2020	2030
1	Receiving ends in eastern and central China	45,934	51,000	47,230
2	Northeast China	8414	8600	830
3	Northwest China	10,752	12,400	1450
4	Sichuan, Chongqing, and Xizang	2676	3200	340
5	Shanxi	5645	6700	710
6	Western Nei Mongol	5719	6700	7300
7	Southern China	11,954	12,954	12,458
Total		91,094	101,554	100,288

5.4.4.1.2 Pumped storage

Under the scenario of low power demand, the construction of pumped storage power stations requires a relatively long time. The pumped storage power stations scheduled to go into operation by 2020 are now basically under construction, so the scale of pumped storage power stations will not change much by that time. By 2030, the installed pumped storage capacity will reach 102.9 GW, up 2.9 GW compared with that under scenario 1. The installed capacities of pumped storage power stations between 2015 and 2030 are shown in Table 5.32.

5.4.4.1.3 Gas power station

Under the scenario of low power demand, the installed capacity of gas power stations will reach 85 GW by 2020, down by 26.5 GW over that under scenario 1. By 2030, the installed capacity of gas power stations will reach 130 billion kW, down by 70 GW over that under scenario 1. The plan for installed gas power capacity in the case of low power demand is shown in Table 5.33.

Table 5.32 The plan on installed capacity of pumped storage power stations (unit: 10,000 kW)

No.	Region	2015	2020	2030
1	Receiving ends in eastern and central China	1392	2142	5372
2	Northeast China	150	35	980
3	Northwest China	0	0	720
4	Sichuan, Chongqing, and Xizang	9	9	340
5	Shanxi	120	12	350
6	Western Nei Mongol	120	12	360
7	Southern China	480	84	2168
Total		2271	3581	10,290

Table 5.33 The plan on installed gas power capacity (unit: 10,000 kW)

No.	Region	2015	2020	2030
1	Receiving ends in eastern and central China	4706	5630	8365
2	Northeast China	34	150	300
3	Northwest China	121	200	400
4	Sichuan, Chongqing, and Xizang	255	300	419
5	Shanxi	268	302	422
6	Western Nei Mongol	0	60	150
7	Southern China	1253	1886	2966
Total		6637	8528	13,022

5.4.4.2 Clean energy consumption
5.4.4.2.1 Overall changes of power flow
Under the demand slowdown scenario, the clean energy development scale will remain unchanged, the scale of cross-regional hydropower and wind power transmission will remain unchanged; under the impact of the reduction of demands are mainly installed coal power capacity and the scale of coal power flow. By 2030, the total scale of coal power flow will decline from 180 to 130 GW under the expected development scenario, down by about 50 GW, as is shown in Table 5.34.

5.4.4.2.2 Clean energy consumption and utilization
As is seen from wind power consumption, compared with scenario 1, the installed wind power capacity will remain unchanged. Against the backdrop of the decline of the load level, the downsizing of the power market, the sharp decline of power transmission corridors at the sending ends, and the little change of the proportion of flexibly adjustable power sources, the proportion of wind curtailment in "Three Norths" regions will increase to a certain extent. On average, the proportion of wind curtailment of China under scenario 4 is higher than that under scenario 1, and will reach about 6.3% in 2030.

Table 5.34 The power transmission demand under the demand slowdown scenario by 2030 (unit: 100 GW)

No.	Region	Coal power	Hydropower	Wind power
1	Receiving ends in eastern and central China	1.3	1	1.3
2	Northeast China	−0.12	–	−0.32
3	Northwest China	−0.42		−0.68
4	Sichuan, Chongqing, and Xizang	0	−1.19[a]	–
5	Shanxi	−0.34	–	–
6	Western Nei Mongol	−0.4	–	−0.3
7	Southern China	–	0.19	–

[a]Including the electricity of 22.4 GW generated from the Three Gorges Hydropower Station.

As is seen from solar energy consumption and utilization, similar to wind power, solar curtailment will also increase, and the proportion of China's solar curtailment will be about 2.1% by 2030.

5.4.4.3 Analysis on typical ways of system operation

5.4.4.3.1 Overall system operation

Considering the cross-regional power transmission in northwest China, in accordance with the requirements on giving priority to renewable energy dispatching and ensuring economical dispatching and operation of the electric power system, an hourly analysis will be given on the output curves and operating positions of various units. The curve of hourly output of various units on a typical day in 2030 is shown in Fig. 5.31.

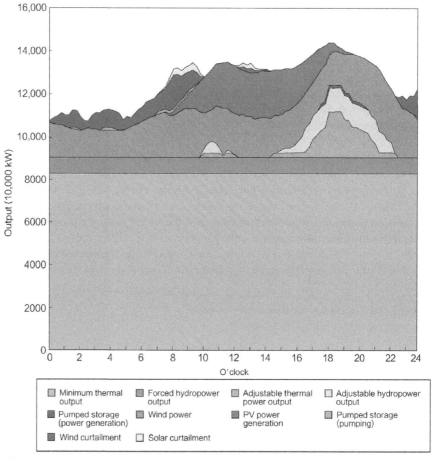

FIG. 5.31

System operation simulation of the power grid on a typical winter day in northwest China in 2030 (including power transmission).

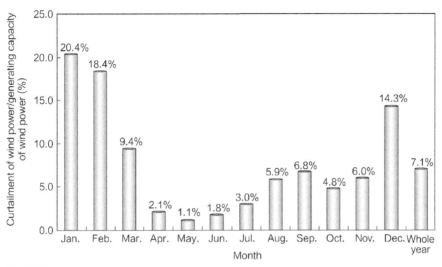

FIG. 5.32

A schematic diagram of wind power consumption and curtailment of the power grid in northwest China in 2030.

5.4.4.3.2 Wind power consumption and utilization

As is seen from wind power consumption and utilization, compared with scenario 1, the installed wind power capacity in northwest China will remain unchanged by 2030. As is seen from the impacts of change factors, since the power demand in northwest China drops by about 10% and the scale of the power transmission corridors reduces, the wind power consumption capability of the power grid in northwest China will drop, and wind curtailment will increase. Research results show that, in the case of the power source structure and load level under scenario 4, the proportion of wind curtailment in northwest China will rise to 7.1%, and will reach about 20% in certain month(s) in winter, as is shown in Fig. 5.32.

5.4.4.3.3 Solar generation consumption and utilization

As is seen from PV consumption and utilization, compared with scenario 1, due to the decline of power demand in northwest China, the PV power-receiving capability of the power grid in northwest China will drop to some extent, and the solar curtailment scale increases. Analysis results show that, in the case of power source structure and load level under scenario 4, the proportion of whole-year solar curtailment will be about 2.9% in northwest China in 2030.

5.4.4.4 Analysis on economical efficiency

A comparison of the economical efficiency is made between scenario 1 (expected development scenario) and scenario 4 (demand slowdown scenario), as is shown in Table 5.35.

Table 5.35 Comparison of economical efficiency under different electric power development scenarios (unit: RMB 100 million yuan)

Economic indicators	Expected development scenario (scenario 1)	Demand slowdown scenario (scenario 4)	Change between scenario 1 and scenario 4
Total cost	310,673	288,347	−22,326
1. Investment	82,343	78,145	−4198
1.1 Wind	12,832	12,832	0
1.2 Nuclear	10,923	10,923	0
1.3 Gas	3052	2686	−366
1.4 Pumped storage	2214	2356	142
1.5 Hydro	9810	9810	0
1.6 Coal	15,547	12,933	−2614
1.7 Other power sources	13,320	13,320	0
1.8 Power grid	14,645	13,285	−1360
2. Fuel cost	184,568	168,319	−16,249
3. Fixed operating expenses	31,558	30,674	−884
4. External environmental expenses	12,204	11,209	−995

Note: *Based on the calculation period of 2011–30, with a discount rate of 8%, investments in the construction of power transmission networks and power distribution networks of the provinces are not covered in the power grid investment.*

We have drawn conclusions on the basis of the calculation results of system optimization:

1) Under the demand slowdown scenario, the system investment will drop by RMB 419.8 billion, the investment in the installed non-fossil energy capacity will remain unchanged, and investments in other power sources than pumped storage and power grid will decline.

2) Under the demand slowdown scenario, due to the reduction of power consumption, the fuel consumption will drop, saving a cost of RMB 1624.9 billion.

3) Under the demand slowdown scenario, due to the reduction of the total installed power capacity, the fixed operating expenses of the electric power system will decline by RMB 88.4 billion.

4) Under the demand slowdown scenario, due to the small fuel consumption and low environmental pollutant emissions, the external environmental expense related to electricity will drop by about RMB 99.5 billion.

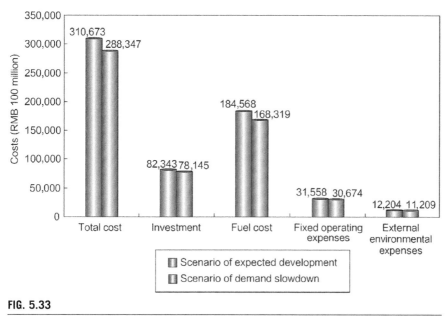

FIG. 5.33

Comparison between the total costs under the demand slowdown scenario and the expected development scenario.

In general, the demand slowdown scenario will drop the total cost of electric power supply by about RMB 2.23 trillion, as is shown in Fig. 5.33.

5.5 COMPREHENSIVE EVALUATIONS OF SCENARIOS

5.5.1 PROPORTION AND COMPOSITION OF NON-FOSSIL ENERGY UNDER DIFFERENT SCENARIOS

5.5.1.1 Proportion of non-fossil energy

Under any of the development scenarios, by 2020, China's non-fossil energy use will account for 15% of its total primary energy consumption, in accordance with the four development scenarios in Section 5.3 and the prediction on the total energy demand in Section 5.2, the estimated proportions of non-fossil energy are shown in Table 5.36.

In the above measurements, non-fossil energy sources for purposes other than power generation are calculated on a full standard (130 million tons of standard coal, see Section 5.2.4). According to current statistical rules, noncommercial energy does not fall into the statistical range of energy consumption. To ensure a consistent statistical standard, the proportions of non-fossil energy sources are measured as per the statistical caliber of commodity energy (the total quantity of other utilized renewable energy sources is 40 million tons of standard coal, see Section 5.2.4), as is shown in

Table 5.36 Proportions of China's non-fossil energy under different scenarios

Scenario	Scenario 1 (hydropower, 350 GW; nuclear power, 60 GW; wind power, 200 GW; solar power, 50 GW)		Scenario 2 (hydropower, 300 GW; nuclear power, 60 GW; wind power, 250 GW; solar power, 100 GW)		Scenario 3 (hydropower, 350 GW; nuclear power, 50 GW; wind power, 210 kW; solar power, 70 GW)		Scenario 4 (hydropower, 350 GW; nuclear power, 60 GW; wind power, 200 GW; solar power, 50 GW)	
	Electricity (100 GWh)	Amount of energy (100 million ton standard coal)	Electricity (100 GWh)	Amount of energy (100 million ton standard coal)	Electricity (100 GWh)	Amount of energy (100 million ton standard coal)	Electricity (100 GWh)	Amount of energy (100 million ton standard coal)
Hydro	12,083	3.69	11,628	3.55	12,083	3.69	12,083	3.69
Nuclear	4368	1.33	4368	1.33	3880.5	1.18	4368	1.33
Wind	3680	1.12	4048	1.23	3680	1.12	3680	1.12
Solar	920	0.28	1104	0.34	1380	0.42	920	0.28
Biomass	680	0.21	680	0.21	680	0.21	520	0.16
Other renewable energy sources	–	1.3	–	1.3	–	1.3	–	1.3
Non-fossil energy total	21,731	7.93	21,828	7.96	21,704	7.92	21,571	7.88
Proportion (total energy demand: 4.5 billion tons of standard coal)	–	17.6%	–	17.7%	–	17.6%	–	17.5%

Proportion (total energy demand, 400 million tons of standard coal)	–	16.5%	–	16.6%	–	16.5%	–	16.4%
Proportion (total energy demand, 5.1 billion tons of standard coal)	–	15.5%	–	15.6%	–	15.5%	–	15.5%

Table 5.37 The proportions of China's non-fossil energy under the scenarios of verification

Scenario of verification	Scenario 1 (%)	Scenario 2 (%)	Scenario 3 (%)	Scenario 4 (%)
Proportion (total energy demand: 4.5 billion tons of standard coal)	15.6	15.7	15.6	15.5
Proportion (total energy demand: 4.8 billion tons of standard coal)	14.6	14.7	14.6	14.5
Proportion (total energy demand: 5.1 billion tons of standard coal)	13.8	13.8	13.8	13.7

Table 5.37, and verifies whether the target of non-fossil energy use accounts for 15% of China's total primary energy consumption.

5.5.1.2 Non-fossil energy mix

According to the above development scenarios, in China's non-fossil energy mix, hydropower will take a leading position and account for at least 40% under any of the four scenarios; nuclear power will account for 14%–17%; the contribution rate of wind power will be 16%–19%, the contribution rate of solar power and biomass power will be relatively small (only 6%–8%), and that of other renewable energy sources will be 16%–17%. As is seen from the overall structure, the non-fossil energy for power generation will account for over 80% of the total scale of non-fossil energy. Under the four development scenarios, the proportions of non-fossil energy sources for achieving the 15% target are shown in Figs. 5.34–5.37.

5.5.2 BENEFITS AND COSTS OF NON-FOSSIL ENERGY DEVELOPMENT UNDER DIFFERENT SCENARIOS

5.5.2.1 Installed power capacity

Assuming that scenario 1 is the baseline scenario, the overall changes of installed power capacity under the four types of scenarios in 2020 and 2030 are shown in Table 5.38 and Fig. 5.38.

5.5.2.2 Clean energy consumption (wind and solar curtailments)

Assuming that scenario 1 is the baseline scenario, the clean energy consumption (wind and solar curtailments) under the four types of scenarios during the planning period of 2011–2030 is shown in Table 5.39 and Fig. 5.39.

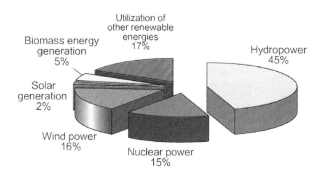

FIG. 5.34

China's non-fossil energy mix of scenario 1 in 2020.

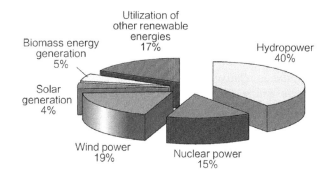

FIG. 5.35

China's non-fossil energy mix of scenario 2 in 2020.

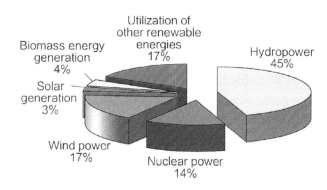

FIG. 5.36

China's non-fossil energy mix of scenario 3 in 2020.

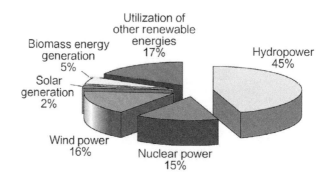

FIG. 5.37

China's non-fossil energy mix of scenario 4 in 2020.

5.5.2.3 Cost of electric power supply

Assuming that scenario 1 is the baseline scenario, the total costs of the electric power supply under the four types of scenarios during the planning period of 2011–2030 are shown in Table 5.40 and Fig. 5.40.

Since the energy density of wind power and solar power is low and the cost of per unit power generation is relatively high, the slowdown of hydropower and nuclear power development will surely increase the installed capacity of the system and the cost of electric power supply (see Tables 5.40 and 5.41). Compared with the baseline scenario, a reduction of about 8% of the installed hydropower capacity, within the planning period, will increase the installed capacity of the system by about 5.0%, and increase the power supply cost by about 3.2%; a reduction of installed nuclear power capacity of about 13% will increase the installed capacity of the system by about 1.8%, and increase the power supply cost by about 1.2%. In addition, if the total power demand is controlled at a relatively low level, the total installed capacity and power supply cost of China will drop significantly. Under the demand slowdown scenario, the power demand will drop by about 11%, the total installed capacity of power source will drop by about 6.2%, and the cost of electric power supply will drop by about 7.3%, compared with that under the baseline scenario.

5.5.3 BASIC CONCLUSIONS OF SCENARIO ANALYSIS

On the basis of the model for overall analysis and evaluation of the non-fossil energy and electric power system proposed in this book, a multiscenario comparative analysis method is used to analyze China's development scale and consumption mode of non-fossil energy under different scenarios, and to demonstrate the techno-economical efficiency, cost effectiveness, and cost. On the basis of the scenario analysis results, we need to focus on solving the following key problems for realizing the target of non-fossil energy use accounting for 15% of the total primary energy consumption.

Table 5.38 Comparison of installed power capacity structures under different scenarios (unit: 10,000 kW)

Type	(Scenario 1) (baseline scenario)		(Scenario 2)-(scenario 1)		(Scenario 3)-(scenario 1)		(Scenario 4)-(scenario 1)	
	2020	2030	2020	2030	2020	2030	2020	2030
Hydro	34,523	43,160	−4857	−3940	0	0	0	0
Pumped storage	3581	10,000	0	2 2	0	13	0	290
Coal	112,400	116,000	600	3	0	4	−10,846	−15,712
Fuel gas	11,152	20,063	300	0	970	1350	−2624	−7041
Nuclear	5824	13,055	0	0	−650	−3 55	0	0
Wind	20,000	40,000	2000	5	0	0	0	0
Biomass	170	3000	0	0	0	0	0	0
Solar	10,000	30,000	2000	5	5000	1	0	0
Total	199,180	275,278	43	838	5320	9995	−13,470	−22,463

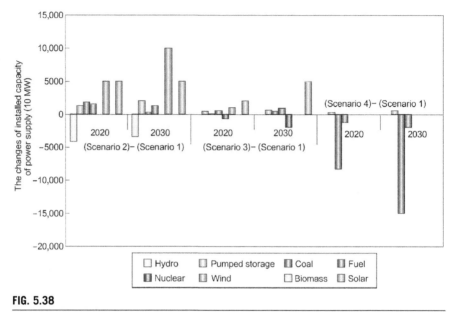

FIG. 5.38

The changes of installed power capacity under different scenarios.

Table 5.39 Proportion of wind and solar curtailments under different scenarios (%)

Type	Scenario 1 (expected development)	Scenario 2 (hydropower slowdown)	Scenario 3 (nuclear power slowdown)	Scenario 4 (demand slowdown)
Proportion of wind curtailment of China	3.5	9.6	4. 1	6.3
Proportion of solar curtailment of China	0.5	3.2	1.5	2.1
Proportion of wind curtailment in northwest China	3.7	11.8	4.8	7.1
Proportion of solar curtailment in northwest China	0.7	5.2	1.4	2.9

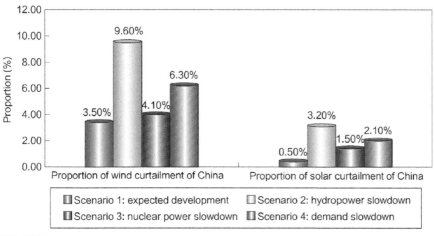

FIG. 5.39

The proportions of wind and solar curtailments under different scenarios.

Table 5.40 Comparison of costs of electric power supply under different scenarios (unit: RMB 100 million yuan)

Type	(Scenario 1) (baseline scenario)	(Scenario 2)-(scenario 1)	(Scenario 3)-(scenario 1)	(Scenario 4)-(scenario 1)
Investment	82,343	7414	2886	−4198
Fuel cost	184,568	1056	142	−16,249
Fixed operating expenses	31,558	1165	767	−884
External environmental expenses	12,204	359	34	−995
Total cost	310,673	9994	3829	−22,326

The basis of achieving the target is to control total energy consumption. The multiscenario analysis results indicate that, if total energy consumption is controlled under 4.8 billion tons of standard coal, the target can be achieved by 2020; if total energy demand grows rapidly, it will reach 5.1 billion tons by 2020, and it will be difficult to achieve the 15% target under the scenario of verification. Therefore, we must attach great importance to energy saving and control the growth of total energy demand. Meanwhile, we may appropriately adjust the statistical standard for energy to incorporate non-fossil energy in noncommercial energy sources in the statistical scope.

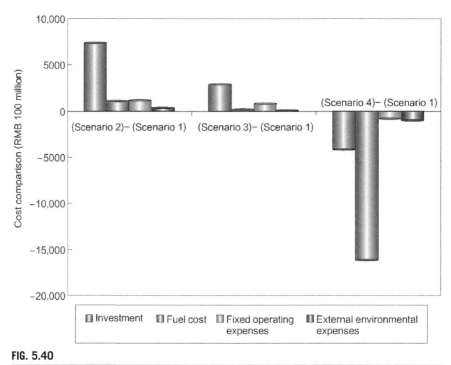

FIG. 5.40

The comparison of electric power supply costs under different scenarios.

Table 5.41 The comparison of costs per kWh with different proportions of installed clean energy capacity

Scenario	Scenario 1 (expected development)	Scenario 2 (hydropower slowdown)	Scenario 3 (nuclear power slowdown)	Scenario 4 (demand slowdown)
Proportion of installed clean energy capacity (%)	34.4	35.5	35.0	36.1
Cost per kWh [yuan/(kWh)]	0.387	0.399	0.392	0.395

Note: *The cost per kWh is the ratio between the present value of the total cost and that of the total energy, the total cost does not contain the investment in power transmission networks and power distribution networks in provinces.*

Achieving the 15% target depends on the utilization of power generation based on non-fossil energy sources. The main way for effective utilization of non-fossil energy is electric power generation. Besides, a small portion of non-fossil energy sources are used directly for heat supply, gas supply, and production fuel, and most non-fossil energy sources are for terminal utilization through power generation, indicating that the power industry plays a central role in achieving the 15% target. It is estimated that, to achieve the target, the proportion of non-fossil energy use to total primary energy consumption needs to reach 12%–13% for power generation, with a contribution rate of more than 80%.

The key to achieving the 15% target is to accelerate the development of hydropower and nuclear power. By 2020, in China's non-fossil energy mix, hydropower will occupy a dominant position, and take up a proportion of more than 40% under any of the scenarios. If the installed hydropower capacity reaches 350 GW and the installed nuclear power capacity reaches 60 GW by 2020, there will be good economical efficiency in achieving the 15% target, due to the small total installed power capacity and the low cost of electric power supply. If the installed hydropower capacity reaches merely 300 GW and nuclear power 50 GW by 2020, to achieve the 15% target of under the scenario of high energy demand, we must accelerate the development of wind power and solar power, achieve higher total installed capacity, and pay higher costs.

The key to accelerating the development of wind power and solar power is to solve the problems of consumption. Wind/solar power feature short construction cycles and are not restricted by project construction cycles. They are mainly restricted by the grid connection and consumption capability of the electric power system. In order to meet the consumption demands of wind power of 200 million–2.5 GW and solar power of 50–100 GW, first, we need to accelerate the construction of cross-regional corridors for power transmission, as scenario analysis results indicate that the proportion of cross-provincial consumption of wind power needs to exceed 50%; second, we shall strengthen the building of the system peak shaving capability, and gradually increase the proportion of power sources of pumped storage and fuel gas that can be flexibly adjusted to increase the total installed power capacity to about 9% by 2020, up by about 4% over that in 2010.

Power grid development and its comprehensive social and economic benefits

6

Energy structural adjustment and non-fossil energy development have put forward new requirements on the functions of China's power grids. In the future, the Chinese government shall fully leverage power grids to optimize and allocate energy resources on a larger scale, solve the long-standing and persistent problems of coal transportation and power transmission, and meet the needs for centralized and distributed development of intermittent power sources including wind power and solar power. For this, the Chinese government needs to accelerate the construction of cross-regional ultrahigh voltage (UHV) power transmissions and receiving-end UHV synchronous power grids, facilitate the building of an integrated system for coal transportation and power transmission, promote long-distance transmission and large-scale balance adjustment of non-fossil energy, meet the needs for large-scale application of new energy and new services of power consumption, facilitate the achievement of the target for non-fossil energy use of 20% accounting for China's total primary energy consumption by 2030, and play a greater role in serving sustainable economic and social development.

6.1 CHINA'S POWER GRID DEVELOPMENT IN THE FUTURE

6.1.1 THE INTERNAL RELATION BETWEEN ENERGY STRUCTURAL ADJUSTMENT AND POWER GRID FUNCTIONS

For a long time, the power grid is simply defined as the carrier of electric energy transmission, and provincial and cross-provincial power grid construction and development are given priority. The existing power grid development model can no longer meet the demands for fulfilling the urgent tasks of energy structural adjustment and non-fossil energy development. As large-capacity and high-efficiency power transmission technologies, such as UHV, become more mature and the electric power system becomes smarter, the power grid will be able to allocate various energy resources more efficiently, and integrate various energy resources more effectively and economically. In the future, the functions of China's power grid will undergo new changes from the four aspects: (1) Meet the needs of large-scale and long-distance transmission of energy and power. (2) Meet the needs of efficient integration and coordinated operation of various energy sources in power-receiving regions.

Non-Fossil Energy Development in China. https://doi.org/10.1016/B978-0-12-813106-0.00006-4

(3) Meet the needs of new energy development and new services for power consumption. (4) Meet the needs of the power market as a physical platform. Looking into the future, the power grid will play a pivotal role in energy development, conversion, transmission, consumption, safety, service, and the market transaction system.

(1) Meet the needs of large-scale and long-distance transmission of energy and power.

As China is still at the stage of rapid development of industrialization, IT application, urbanization, and agricultural modernization, its power demand will continue to rise, and its eastern and central regions will serve as the major power load center of China for a long time to come. Except for nuclear power, newly developed coal power, hydropower, wind power, and solar power mainly concentrate in Western and Northern China. Power demand and the supply of power-generating energy show a characteristic of reverse distribution, determining that large-scale and long-distance energy delivery is an important feature of China's mid- and long-term energy structural adjustment and energy pattern evolution. The power flow of the "West-East Power Transmission" and "North-South Power Transmission" projects will further increase the hydropower of Southwest China, the coal power of Western China and Northern China, the wind power of the "Three Norths region," and the solar power of Western China will be delivered to the eastern and central regions of China on a large scale, and China's neighboring countries including Russia, Mongolia, and countries in Central Asia and Southeast Asia will deliver electricity to China's load centers. As an important part of China's integrated energy transport system, the power grid will play an important role in coal power optimization, large-scale allocation, and long-distance transmission of non-fossil clean energy, large-scale power transmission, and the building of an integrated energy transportation system in China. The energy and power flows are shown in Fig. 6.1.

(2) Meet the needs of efficient integration and coordinated operation of various energy sources in power-receiving regions.

In the future, as the cross-regional power flow increases over time, the power-receiving regions including Beijing, Tianjin, Hebei, and Shandong, the four provinces in Central China and the five provinces (autonomous regions) in East China will receive more electricity, the electric power system operation will become more complicated, and nonlocal coal power, hydropower, and new energy power will be used, together with local coal power, gas power, hydropower, pumped storage, nuclear power, and new energy, in a coordinated way, putting forward higher requirements on the resource allocation capability and coordinated operation of the receiving-end power grid, namely larger coverage of the receiving-end power grid, a more robust grid, higher dynamic balance capability, and increased safety and stability.

FIG. 6.1

A schematic map of integrated national energy bases in China.

From: National Energy Administration, the "12th Five-year Plan" on Energy Development in China.

(3) Meet the needs of new energy development and new services of electricity consumption.

   Power generation based on new energy sources such as wind energy and solar energy are characterized by randomness and intermittency, and its controllability and predictability are lower than that of power generation based on traditional fossil energy. Large-scale development and utilization has brought enormous challenges to the control and coordination capabilities of power grid, and requires the use of advanced automation technology, coordination control technology, and energy storage technology in the power grid, so as to realize precise control and efficient utilization of new energy. Meanwhile, social advancements and the development of new electrical applications such as the electric car and smart home appliances have put forward higher requirements on electric energy supply quality and electric service contents, and smarter application will become a major trend of the development of the power grid.

(4) Meet the needs of establishing a unified power market in China.

The operation of China's power market platform needs to be based on a countrywide power grid for addressing the fundamental role of the market in the allocation of energy resources to a greater extent and at a wider range, promote large-scale development and efficient utilization of non-fossil energy, and gain greater benefits in terms of economy, energy saving, resources, environment, safety, and social development. In order to adapt to the functional development and changes of China's power grids in the future, the power grid must be both "robust" and "smart." As China keeps making a significant breakthrough in UHV power transmission, smart grid construction is also under way, and thus a smart grid will form with UHV at its core and coordinated power grid development at different levels.

To sum up, in the future, China's energy development will be characterized by increasing concentration of energy production and consumption, intensified reverse distribution of energy production and consumption, and higher demands for cross-regional energy dispatching, requiring the Chinese government to change the energy allocation manner of excessive reliance on coal transportation, speed the development of cross-regional power transmission, place equal emphasis on coal transportation and power transmission, promote smart grids, propel the transformation of China's energy structural adjustment and transformation of energy development model, and realize energy development in a safe, efficient, and clean way. The internal relation between energy structural adjustment and power grid function is shown in Fig. 6.2.

6.1.2 THE OVERALL SITUATION OF CHINA'S POWER FLOW

In accordance with the characteristics of the distribution of China's energy resources and for the purpose of sustainable development of energy and power, China will give priority to the construction of large coal power bases, large hydropower bases, large nuclear power bases, and large new energy bases in its resource-rich regions and load centers, so as to meet the power needs of its economic and social development. Geographically, in the future, China's large nuclear power bases will be mainly built in coastal areas and energy-hungry inland provinces, close to load centers for local consumption. However, the large coal power bases in Northern and Western China, the large hydropower bases in Southwest China, the wind power bases in Northern and Western China, and the solar power generation bases in Western China are far away from load centers, and the electricity generated in such regions needs to be delivered to load centers including Beijing, Tianjin, Hebei, and Shandong, the four provinces in Central and Eastern China, the five provinces in East China as well as Guangdong and Guangxi in Southern China. The overall situation of China's power flow in the future is shown in Fig. 6.3.

By giving overall consideration to China's distribution of power loads and power sources, the power flow of "West-East Power Transmission" and "North-South

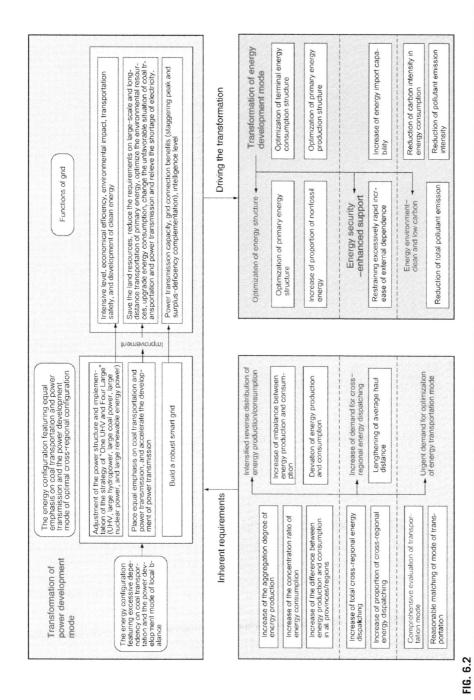

FIG. 6.2

The internal relationship between energy structural adjustment and power grid function.

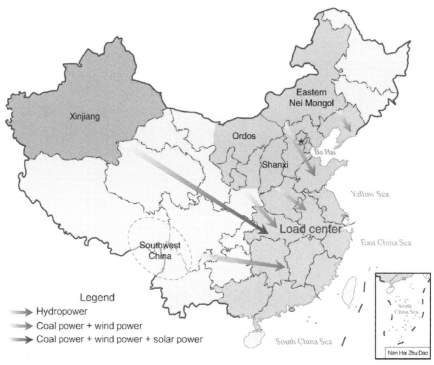

FIG. 6.3

The overall situation of China's power flow in the future.

Power Transmission" will be further expanded. Currently, the eastern and central regions of China receive a power flow of 110 GW, which is expected to rise to about 500 GW in 2030.

6.1.3 CHINA'S POWER GRID DEVELOPMENT IN THE FUTURE

The development of China's power grids shall be in line with regional planning and the urban planning of its 19 city clusters, and be adapted to the basic situation of the centers of energy resources and energy consumption of the country. In the future, China needs to address the problems of inadequate capability of cross-regional power transmission and low power supply reliability of power distribution networks, rely on the UHV power grid, and focus on the development of large-capacity, long-distance, and cross-regional power transmission corridors; the Chinese government shall energetically strengthen the construction of power distribution networks, accelerate to improve the urban power grid, make steady progress in power grid construction and renovation in rural areas; make overall plans for the development of power grids at different levels, and form a robust power grid with reasonable grid structure and strong capability of resources allocation.

After years of demonstrations, we have drawn the following conclusions: in the future, China will form an interconnected direct current (DC) system consisting of the four synchronous power grids of North China-Central China-East China, Northeast China, Northwest China, and Southern China. Specifically, the "North China-Central China-East China" power grid has a "cluster" structure, is supported by a 1000-kV receiving-end grid, has a closer electrical connection, and will create favorable conditions for receiving electricity from energy bases on a large scale. The hydropower of Southwest China, coal power of Western and Northern China, wind power of "Three Norths," and solar power of Western China will be delivered to the "North China, East China, and Central China" power grid load centers, via cross-regional power grids, and the power flow of the "West-East Power Transmission" and "North-South Power Transmission" projects will increase significantly, as shown in Fig. 6.4. Meanwhile, China will interconnect with Russia, Mongolia, and countries in Central Asia and Southeast Asia, so as to achieve power grid interconnection and optimize the allocation of resources in a greater range.

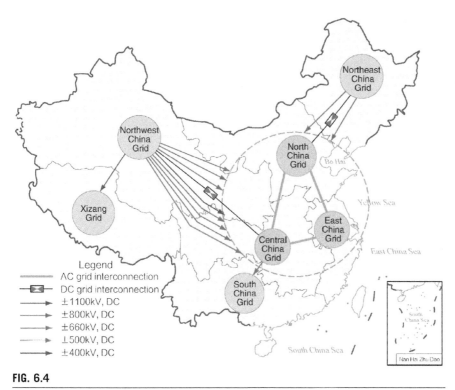

FIG. 6.4

The distribution of synchronous power grids in China.

6.2 THE DRIVING EFFECTS OF POWER GRID ON NON-FOSSIL ENERGY DEVELOPMENT

6.2.1 PROMOTE HYDROPOWER DEVELOPMENT AND TRANSMISSION AND EFFICIENT UTILIZATION

(1) Promote the development of hydropower base and power transmission.

In spite of its huge reserves of hydropower resources, Southwest China has a low level of hydropower development and the situation needs to be improved in the future. It is estimated that the installed hydropower capacity will increase by more than 130 GW in the next 15 years. As is seen from the hydropower consumption market, Sichuan, Yunnan, and Xizang are less developed and have a relatively low power demand level. According to the prediction of scenario 1, by 2020, the power consumption of the three provinces will reach 470 TWh, with a maximum load of 92 GW. Therefore, huge technical available hydropower resources need to be delivered out of Southwest China via a long-distance power transmission on a large scale.

According to the predictions on hydropower development planning and the power demand of Southwest China, hydropower generated from the region needs to be delivered to load centers in Eastern and Central China on a large scale. The hydropower base in Southwest China is 1000–3000 km from the load centers in Eastern and Central China, and for this, it is necessary to build UHV alternating current/direct current (AC/DC) power transmission corridors for delivering the rich electric energy resources in Southwest China to the load centers in Eastern and Central China where there is a huge market demand, and then distribute and utilize the resources via a large power grid. It is predicted that by 2020 the installed capacity for hydropower transmission from Southwest China will reach 63 GW, accounting for 18% of the total installed hydropower capacity of China, and the electricity for transmission will reach 250 TWh, accounting for 21% of the total generated hydropower of China. By 2030, the installed capacity for hydropower transmission from Southwest China via a UHV AC/DC power grid is expected to reach 120 GW. Specifically, the Xiangjiaba, Xi Luodu, Baihetan, and Wudongde large hydropower stations at the lower reaches of the Jinsha River are located at the border between Sichuan and Yunnan, far away from local load centers compared with other local hydropower stations, and for this to be the best choice for outward power transmission, the model of ± 800-kV DC power transmission will be adopted; the hydropower stations (Jinping Stage-I and Jinping Stage-II) on the lower reach of Yalong River, featuring a large scale, concentrated capacity in the startup period, and long distance of power transmission, will also adopt the model of ± 800-kV DC power transmission; on a long-term basis, it is also appropriate to use ± 800-kV DC or ± 1100-kV DC power transmission for the hydropower resources of Xizang. Meanwhile, for other hydropower resources within Sichuan, the model of

1000-kV AC may be used for outward hydropower transmission and create favorable conditions for strong DC and strong AC parallel operations.

In addition, the provinces in Southwest China, including Sichuan, have an energy structure in which hydroenergy plays a dominant role but fossil energy sources, such as coal, are lacking. Moreover, the precipitation in Sichuan has periodic and seasonal characteristics, and the proportion of hydropower stations with a long-term regulating performance is not high, causing a very uneven distribution of hydropower resources. It is an effective way to ensure electric power supply in Sichuan during both the flood season and dry season and to build a safe, stable, and an economically efficient energy supply system by delivering thermal power from Northwest China (mainly coal power from Xinjiang) to Sichuan, and establishing a coordinated system of thermal and hydropower operations. As a result of the large-scale hydropower development in Xizang, part of the hydropower from Xizang can be delivered to the power grid of Sichuan to meet the power demands of the province and achieve sustainable utilization of the outward hydropower transmission corridors. The completion of the construction of the "North China, East China, and Central China" UHV synchronous power grid and the DC power grids in Southwest China and Northwest China can ensure large-scale electric power transmissions in Southwest China, including Sichuan, during different seasons and periods, and achieve thermal-hydropower complementation and power regulation between flood and dry seasons within a larger range.

(2) Reduce water curtailment and increase hydropower utilization efficiency.

At present, hydropower forms a large proportion of the electric power system in Southwest China, and small hydropower stations take a high proportion in hydropower stations. The precipitation of Southwest China has clear seasonal characteristics, the Yangtze River basin has a concentrated runoff, while the small hydropower stations have very poor regulating capability, and thus there is a serious problem of curtailment of water for power generation in Sichuan and Chongqing in summer. Along with the further development of hydropower in Southwest China, on the one hand, the construction of outward hydropower transmission corridors in Southwest China can not only ensure large-scale power transmission from hydropower stations on the Jinsha River and Yalong River, but also solve the problem of curtailment of hydropower generation in Sichuan in summer; on the other hand, the establishment of "North China, East China, and Central China" UHV synchronous power grids may further expand the market of the hydropower from Southwest China, and realize efficient consumption of hydropower from the southwest region to the eastern and central regions of China through cross-regional regulation. Research results show that, through the construction of the hydropower transmission corridors in southwest China and the "North China, East China, and Central China" UHV synchronous power grid, it is expected to reduce the hydropower curtailment each year by 2020 by over 20 TWh, and to achieve non-fossil energy development goals.

6.2.2 PROMOTE THE LARGE-SCALE DEVELOPMENT AND EFFICIENT CONSUMPTION OF NEW ENERGY

(1) Promote centralized development and long-distance transmission of wind power.

Main wind power bases in China's "Three Norths" region arc restricted by such factors as local power demand and scale of the power grid, and have a limited wind power consumption capability. In order to promote further development and utilization of wind power, it is necessary to connect the wind power to the regional power grids and even to the power grids outside the region. According to the priority of wind power consumption, the consumption markets of the main wind power bases in China's "Three Norths" region are presented below.

In Northeast China, Liaoning, Jilin, Heilongjiang, and Eastern Nei Mongol are key regions of China's wind power development. Specifically, Liaoning hardly caters to nonlocal wind power, since the wind power it needs can be supplied locally. There is a small market potential for expanding wind power consumption in Heilongjiang, Jilin, Eastern Nei Mongol, and the large-scale wind power development there needs to rely on cross-regional transmission. In Northwest China, the wind power of Gansu is used for local consumption, delivered to the regional power grid through the hydropower regulation capability on the upper reach of the Yellow River in Qinghai, and delivered out of the region via UHV DC cross-regional transmission. Since the power grid of Northwest China is capable of consuming the wind power generated at the Jiuquan wind power base, the wind power generated in Ningxia and Xinjiang needs to be delivered out of the region, via cross-regional transmission corridors, to meet the demands for local consumption. In the "North China, East China, and Central China" region, the wind power generated in Western Nei Mongol will be used in the Nei Mongol and Beijing-Tianjin-Tangshan regions and delivered to the load centers in Eastern and Central China. The wind power generated in Hebei will be used in Beijing, Tianjin, and Hebei, and used for the receiving-end load centers in the "North China, East China, and Central China" region. The wind power generated in Shanxi, Shandong, and Jiangsu needs to be consumed and utilized in the "North China, East China, and Central China" region. The wind power consumption of China in the future is shown in Fig. 6.5.

According to the "13th Five-year Plan" on wind power development, China's wind power development is greatly different from that during the 12th Five-year Plan period, and the situation that more than 50% of the wind power is used for outward power transmission has changed. According to the "13th Five-year Plan" on renewable energy development, by 2020, China's installed wind power capacity is expected to reach 210 GW. As is seen from the distribution of wind power stations, it is hard to change the situation that China's wind power development in the "Three Norths" region plays a leading

FIG. 6.5

The wind power consumption of China in the future.

role on a long-term basis, and that the installed wind power capacity in the "Three Norths" region will account for about 65% of that in China. As is seen from the wind power consumption market, according to the analysis results of scenario 1, by 2020 China's installed capacity for cross-provincial cross-regional wind power transmission will exceed 40 GW, accounting for about 20% of the total installed wind power capacity of China. The schematic diagram of China's wind power consumption by 2020 is presented in Fig. 6.6.

(2) The effects of the "North China, East China, and Central China" UHV synchronous power grid on promoting wind power development.

The eastern and central regions of China need to efficiently consume not only the electricity from the planned wind power bases in Hebei, Shandong, and Jiangsu, but also wind power from the wind power bases in the Northeast China and Northwest China, putting forward very high demands on the regulating capability of the power grid of Eastern and Central China. The "North China, East China, and Central China" UHV synchronous power grid may help to build a bigger platform for optimization and allocation of resources, and solve the problem of inadequate capability of large-scale wind power development and consumption in China.

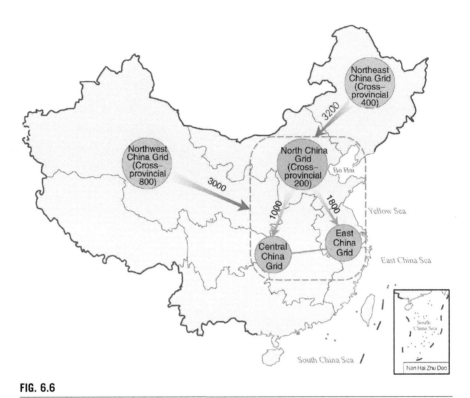

FIG. 6.6

A schematic diagram of China's wind energy consumption in 2020. Note: the unit of the data in the figure is GW.

In North China, East China, and Central China, there are certain seasonal and time differences, and the maximum and minimum system loads occur in different months and period. The construction of a unified "North China, East China, and Central China" UHV synchronous power grid may bring obvious staggered peaks, lower the peak-valley difference, and help to increase the wind power consumption capability of the system. The fitting load curves of "North China, East China, and Central China" power grid on a typical winter day and a typical summer day are shown in Figs. 6.7 and 6.8, respectively. It is clear that the "North China, East China, and Central China" power grid will lower the maximum load of the system, reduce the difference between peak and valley, and stimulate wind power consumption noticeably.

As is seen from the power grids in the eastern and central regions of China, due to the great differences between the power source structures, it is relatively difficult for the power grids in North China and East China to consume wind power; while it is relatively difficult for the power grid of Central China to consume hydropower in summer, in the winter, the

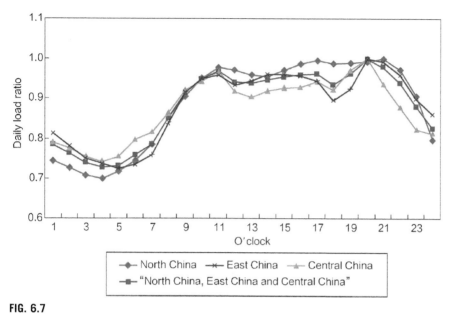

FIG. 6.7

The load curve of "North China, East China, and Central China" power grid on a typical summer day.

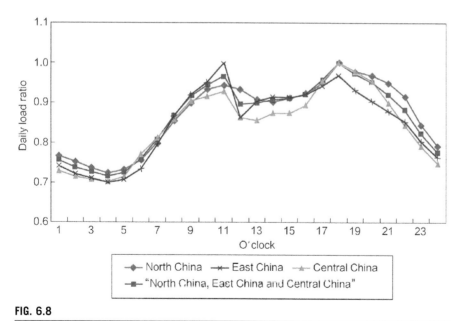

FIG. 6.8

The load curve of "North China, East China, and Central China" power grid on a typical winter day.

hydropower from Central China may be used for system peak shaving. Therefore, after the construction of the "North China, East China, and Central China" UHV synchronous power grid, the regulating capability of the hydropower stations in Central China may be fully leveraged to stimulate wind power consumption in North China and East China at the time of difficult wind power consumption in winter.

Research results indicate that, at the acceptable level of wind curtailment (5%), the construction of the "North China, East China, and Central China" UHV synchronous power grid (including Western Nei Mongol), may help increase the wind power consumption capability to about 120 GW by 2020, an increase of about 40 GW. The "North China, East China, and Central China" UHV synchronous power grid not only meets the demands for wind power consumption in the region, but also provides a consumption platform for wind power from Northeast China and Northwest China.

If the construction of cross-regional power grids is not considered on the existing basis, plus the established target of installed wind power capacity, a very serious problem of wind curtailment may occur at that time. Currently, China's "Three Norths" area has seen very severe wind curtailment due to the backward cross-regional power grids and limited wind power consumption market. In 2015, the wind curtailment ratio in Gansu, Xinjiang, and Jilin exceeded 30%. The increasing installed wind power capacity under the condition of backward power grid construction will worsen the problem of wind curtailment in the future. Through preliminary analyses, if the construction of a cross-regional power grid is not considered on the existing basis, it is estimated that the wind power curtailment will reach 68 TWh by 2020, equivalent to 22 million tons of standard coal. Assuming that China's primary energy consumption will reach 4.8–5.1 billion tons of standard coal by 2020, it will cause the proportion of China's non-fossil energy to decline by about 0.4 percentage points by 2020.

Nei Mongol, Gansu, and Xinjiang are very rich in solar energy resources. On the basis of local economic and geographical conditions, it is reasonable to utilize solar power by building large grid-connected solar PV power bases in deserts, Gobi, and wastelands with the richest solar resources for large-scale and centralized solar development. However, similar to wind power, solar generation is also intermittent, and large-scale solar generation and grid connection would propose higher requirements on safety and stability control of the electric power system. In the future, China's solar energy PV power generation bases will be mainly located in remote areas in Western China, where local power grids can only receive limited solar energy, and where solar PV power generation bases need large-capacity and long-distance power transmission. In order to promote large-scale and intensive development and utilization of solar power from places that are far away from load centers, we need to build cross-regional power grids, achieve coordinated development, integrated operation, and cross-regional transmission of wind power, solar

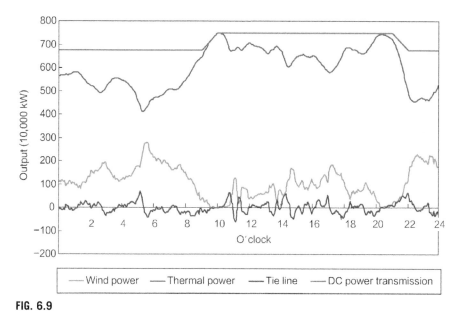

FIG. 6.9

Joint delivery of renewable energy and coal power.

power, hydropower, and thermal power, and solve the economic and technical problems in the process of large-scale and long-distance transmission of generated electricity based on clean energy. The joint delivery of renewable energy and coal power is shown in Fig. 6.9.

(3) Promote distributed development and the in situ utilization of new energy.

The Chinese government places equal emphasis on the development of new energy. Besides centralized and large-scale development, it shall build distributed new energy power stations of appropriate size in load centers in Eastern and Central China, on the basis of the specific conditions of wind and solar energy resources, and promote efficient utilization of distributed new energy through strengthening construction and renovation of the power distribution network. For wind power, in Central China, Southern China, and Eastern China, the specific conditions of wind energy resources shall be considered so as to make the best of the regulating capability of the local power grid, and achieve distributed development and local consumption of wind power, according to local conditions. For solar generation, first, the Chinese government shall address the advantages of distributed solar power supply, promote household PV power generation systems or build small PV power stations in remote areas and islands such as Xizang, Qinghai, Nei Mongol, Xinjiang, Ningxia, Gansu, and Yunnan, so as to ensure power supplies to such areas; second, build rooftop solar grid-connected PV power generation facilities and increase the urban use of renewable energy in Beijing, Shanghai, and in economically developed large- and medium-sized cities in Jiangsu, Zhejiang, and Guangdong.

Wind power and PV power are random and intermittent, and have basically no capacity benefits. Distributed wind power and PV power generation have low-voltage grade, but feature great output fluctuation and low-capacity benefits. Therefore, conventional installed power capacity should not be replaced by the distributed development of new energy, and a large power grid needs to provide users with enough spare capacity. As is seen from the total system cost, except for the investment in new distributed power stations, the investments in large power grids do not decline, the utilization rate of large power grids drops, and the development of distributed power generation will raise the total system costs. End users shall bear the investment costs of distributed power generation, and the capacity reserve cost of large power grids; if subsidies to distributed power generation is not considered, the electric charge payable by relevant electricity users will increase.

The smart grid plays an important role in meeting the requirements on distributed grid connection and local consumption of new energy. By integrating advanced information, automation, energy storage, operational control, and dispatching technologies, a smart grid is capable of predicting the output of electricity generated from various new energy sources, ensuring the optimal operation of various power units, improving the output characteristics of new energy power generation, solving the problems related to power grid safety and stability caused by high-proportional grid connection of new energy, effectively enhancing the capability of power grid for receiving electricity generated from new energy, and facilitating sustainable development and consumption of new energy.

The panorama of the smart grid is shown in Fig. 6.10.

6.2.3 PROMOTE SAFE AND ECONOMICAL OPERATION OF NUCLEAR POWER

China gives priority to the nuclear power development in its eastern coastal areas and energy-hungry central region. The installed nuclear power capacity of the eastern coastal areas and Northeast China will reach about 30 GW by 2020. In the next 10–20 years, the Chinese government will focus on the development of large-capacity nuclear power plants. Since there are relatively high criteria for the selection of nuclear power plant sites, plus the shortage of land in the eastern and central regions of China and the consideration of intensive nuclear power development, the number of planned units may range from six to eight in the future, and such units will be planned once and built in different stages. As is seen from the capacity of a nuclear power unit, in the future, third-generation nuclear power technology will be mainly used in China for nuclear power units, which has a unit capacity of over 1 GW (that of the nuclear power unit of the AP1000 technology of the United States is

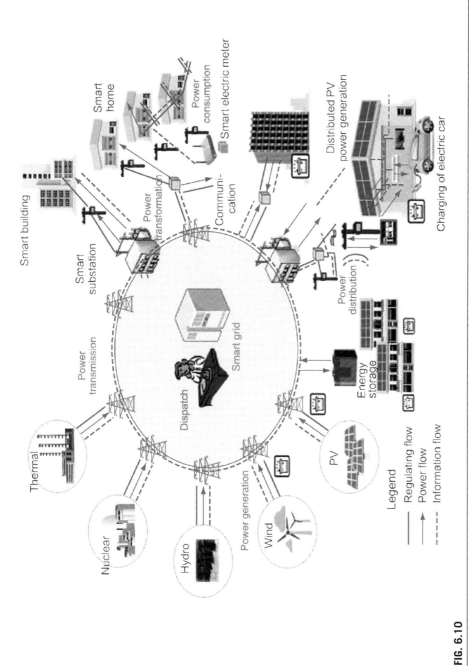

FIG. 6.10

A schematic panorama of the smart grid.

1.25 GW, and that of the nuclear power unit of the French Evolutionary Power Reactor (EPR) technology is 1.75 GW). Therefore, in the future, the installed capacity of most of China's nuclear power stations may ultimately reach at least 8 GW.

The large-scale development of nuclear power has placed higher requirements on the adaptability of the power grid, and UHV power transmission technology will play an important role in promoting the development of large nuclear power bases. On the one hand, as the number of nuclear power stations in the receiving-end system (particularly in the coastal areas of East China) increases, the power transmission via a 500-kV power grid will make it much more difficult to choose a corridor, cause unreasonable technological economy, and make it more difficult to solve the problem of out-of-limit system short-circuit current; on the other hand, as the DC power flow to the coastal areas in East China keeps increasing, relying on or enhancing the current 500-kV power grid alone may give rise to large-area blackouts due to the failure in continuous DC phase changes. In the case of major failure, the nuclear power units in East China may have problems of voltage instability, and for this reason, cannot ensure stable operation of nuclear power units. Therefore, the construction of the "North China, East China, and Central China" UHV synchronous power grid plays an important role in effectively reducing the safety risks arising from intensive 10-plus-loop DC feeding into the power grid of East China and ensuring safe nuclear power operation and efficient evacuation in the receiving-end regions. In addition, under the circumstances of large-scale nuclear power development in the eastern and central regions of China, it is necessary to accelerate the construction of local peak shaving power sources, since nuclear power is basically not involved in peak shaving. Meanwhile, it is necessary to make use of the UHV power grid and the surplus regulating capability of other regional power grids (including the hydropower regulation capability of Southwest China during the dry season), thereby ensuring that the nuclear power stations in coastal areas can operate smoothly under the rated conditions, and ensuring safety and economical efficiency of nuclear power operations.

6.2.4 COMPREHENSIVE EVALUATION OF THE EFFECTS OF POWER GRIDS ON PROMOTING NON-FOSSIL ENERGY DEVELOPMENT

The distinctive features of China's power grid development in the future are strong transmission capability and efficient allocation. Power grids will play a pivotal role in promoting non-fossil energy development, as shown in Fig. 6.11.

Among China's power stations based on various non-fossil energy sources, according to the analysis results in scenario 1, by 2020, the installed capacity of China's cross-regional power grid for outward hydropower transmission will reach 63 GW, accounting for 21% of the total hydropower generation of China; the installed capacity of cross-regional power grids for wind power transmission will reach 40 GW, accounting for about 20% of the total wind power generation of the country; "North China, East China, and Central China" UHV synchronous power grid will receive nuclear power of 29.4 GW, accounting for about 50% of the total nuclear power generation of the country. In addition, as solar generation develops on

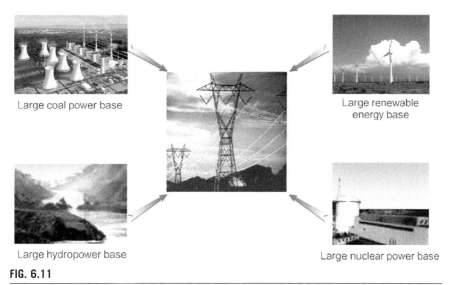

Large coal power base

Large renewable energy base

Large hydropower base

Large nuclear power base

FIG. 6.11

The pivotal role of power grids.

a large scale, the solar generation bases in Western China may also need outward power transmission. Research results indicate that, by 2020, China's cross-regional power grids will be capable of transmitting and allocating electricity generated from non-fossil energy sources of about 600 TWh, equivalent to 180 million tons of standard coal, with a rate of about 25% of contribution to the target of non-fossil energy use accounting for 15% of the total primary energy use by 2020.

6.3 COMPREHENSIVE SOCIAL AND ECONOMIC BENEFITS OF POWER GRID DEVELOPMENT

In the future, China's power grid construction needs to meet the needs of energy structural adjustment, optimization of power source distribution, centralized and distributed development of non-fossil energy, diversified user services, etc. For this, on the one hand, China needs to strengthen the construction of its main grids, promote the coordinated development of power grids at different levels, and raise its capability of cross-regional and large-scale resource allocation through power grids; on the other hand, it needs to base the power grid more on IT applications, automation and interaction, and make the power grid smarter. Building a more robust and smarter power grid will make an important impact on China's energy, environment, economy, and society. This section, based on the evaluation methods for energy development scenarios and benefits mentioned in Section 3.3, analyzes the comprehensive social and economic benefits of power grid development by using the GCE model and the input-output (IO) model.

6.3.1 ECONOMIC BENEFITS

(1) Increase the economical efficiency of energy delivery, and reduce the cost of electric power supply.

Considering the conditions of coal transportation and the latest development of power transmission technology, for coal transportation and power transmission from coal power bases in Western and Northern China to the load centers in Eastern and Central China, power transmission is more economical than coal transportation regardless of the perspective of electricity price or the price at the process of power transmission. The electricity price in the power-receiving regions is RMB 0.06–0.11/(kWh) lower than the on-grid price of coal transportation to power-receiving regions for building coal-fired power plants. The construction of cross-regional power grids and increasing the proportion of power transmission from energy bases in Western and Northern China may effectively increase the economical efficiency of energy delivery and reduce power supply costs. In the medium and long term, the installed capacity for power generation from major coal-producing areas to load centers in Eastern and Central China is expected to be 150 GW, equivalent to air transportation of 220 million tons of coal, saving a total cost of electric power supply of about RMB 60 billion/year.

(2) Increase the utilization efficiency of power grid equipment, and reduce the investment in power grid construction.

The smart grid fully leverages various resources at both sides of supply and demand, to ensure economical system operation, effective management of the loads of power transmission and distribution facilities, and postpone and reduce the construction of power grid facilities. Analysis results show that, after China builds its smart grid in 2020, China's power loads will drop by about 49 GW compared with conventional power grids. Assuming that the investment in power grid equipment is RMB 3700 yuan/kVA, it may reduce such investment by about RMB 18 billion.

(3) Address grid connection benefits, and reduce effective installed capacity and power source investment.

The construction of the "North China, East China, and Central China" UHV synchronous power grid may help give full play to the benefits of peak staggering and peak shaving, and reduce peak load, peak-valley differences of power grid load and the reserve ratio. According to the statistical data of the past decade, the maximum load coincidence factor of the North China, East China, and Central China power grid is 0.968. According to the prediction on energy power demands, by 2020 the maximum load of the power grids in North China, East China, and Central China is expected to reach about 750 GW, and the peak staggering benefits of the "North China, East China, and Central China" UHV synchronous power grid is expected to reach 24 GW. If the reserve ratio is considered, it may reduce the effective installed capacity of 27 GW required by electric balance and lower the power source investment by about RMB 100 billion.

6.3.2 **ENERGY-SAVING BENEFITS**

(1) Improve the comprehensive development and utilization efficiency of coal by means of rational coal transportation and power transmission.

Currently, the coal of China's coal-producing areas is mainly delivered via railway and highway and by sea if the coal is to be delivered to load centers in Eastern and Central China. In the case of joint railway-sea coal transportation, it involves railway transportation, sea transportation, and port operations, resulting in great losses during transportation; in addition, the power coal from the "Three Westerns" region (Western Shanxi, Western Shaanxi, and Western Nei Mongol) is of poor quality, with an average calorific value of about 4700 kcal/kg. Compared with the transportation of cleaned coal, it wastes a railway transport capacity of about 200 million tons, and is a major cause of the current shortage of coal power. Due to the shortage of railway transportation, much coal has been moved by highway transportation. In 2011, about 400 million tons of coal was transported from the "Three Western" region via highways, which causes high energy consumption, and does not comply with China's national conditions of inadequate capability of petroleum supply, since highway transportation means transportation of low-level energy (petroleum) by means of high-level energy (coal). Through rational coal transportation and power transmission, cleaned coal with a higher calorific value is transported to the eastern and central regions of China, but intermediate coal, coal gangue, raw coal (including lignite) with a relatively low calorific value is converted locally into electricity, which is also delivered to the eastern and central regions of China. Meanwhile, it can reduce the scale of coal transportation and energy transportation loss. By 2020, the highway coal transportation is expected to drop to less than 100 million tons, and the efficiency of the railway transportation of cleaned coal will rise, thereby increasing the efficiency of coal transportation and consumption to the greatest extent. Although the coal consumption is relatively high, since air-cooled units are used at the sending ends for power generation, as seen from the whole process of coal development, conversion, transportation, and utilization, in the case of placing equal emphasis on coal transportation and power transmission, the overall efficiency of energy utilization is still 0.2–0.3 percentage points higher than that of direct transportation of coal. By means of the construction of cross-regional power grids, it is estimated that 16.5 million tons of fuel for highway and sea transportation and electric power of 1.2 TWh for electric locomotives in railway transportation may be saved by 2020.

(2) Reduce fossil energy consumption and safeguard China's energy supply security.

Most clean energy sources need to be converted into electricity for grid connection, thus the power grid plays an important role in promoting the development and utilization of clean energy. Taking scenario 1 for an example, by 2020 China's installed hydropower capacity will reach 345 GW, nuclear power 58 GW, wind power 200 GW, solar power 100 GW, and biomass power 15 GW. It is estimated that China's electricity, generated from clean energy will reach 2.3 trillion kWh, equivalent to about 700 million tons of standard coal, and will reduce corresponding fossil energy consumption by 2020. Assuming that China's total energy consumption is 5.1 billion tons of standard coal by 2020, the capability of sustainable supply of fossil energy will be 3.6 billion tons of standard coal and the total fossil energy supply will reach 1.5 billion tons. Therefore, clean energy can make up about 50% of the energy supply-demand gap and play an important role in safeguarding China's energy supply security.

(3) Increase operation efficiency of the electric power system, and reduce coal consumption for power generation.

By building a platform for friendly interactions between power source, power grid, and users, the smart grid, under the mechanism of time-of-use electricity price and so on, guides the users to shift power loads during the peak period to the valley period and reduce the difference between peak and valley. Moreover, it uses peak shaving power sources effectively, so as to reduce the frequency and range of output regulation of thermal power units, increase the efficiency of thermal power units, reduce component wear, and increase operating economical efficiency. It is estimated that the smart grid will make the unit coal consumption of the thermal power station for power generation drop by 4–6 g/(kWh) by 2020.

(4) Increase power grid transmission efficiency and lower line loss.

It will significantly reduce the power loss in the process of electric energy transmission by building a robust power grid with a long-distance, large-capacity, and low-loss UHV power grid, which serves as a backbone, and power grids at different levels develop in a coordinated way; the control of smart scheduling systems and flexible power transmission technologies, over smart sites and real time, and two-way interactions with power consumers, can also optimize the flow distribution of the system, and enhance the transmission efficiency of the power transmission network; meanwhile, the construction of the smart grid will facilitate the extensive use of distributed energy, and reduce to a certain extent the losses caused by power transmission. By estimate, a line loss of about 7.2 TWh may be reduced by 2020.

(5) Increase energy utilization efficiency of terminal electric equipment.

Through a smart interactive terminal, users may, based on their habits of power use, electricity price level, and electricity-consuming environment, set parameters for smart home appliances, thereby increasing the electric utilization efficiency of terminal equipment and saving electric power. To estimate, by 2020 the smart grid may reduce power consumption by about 44.5 TWh and save fuel costs of about RMB 9.63 billion (It is assumed that on average per 1 kWh of electricity will consume fuel with a cost of about RMB 0.216).

6.3.3 RESOURCE AND ENVIRONMENTAL BENEFITS

(1)   Optimize the utilization of China's environmental resources, and reduce environmental loss.

China's maximum allowance of sulfuric deposition shows a tendency of "low in the east and high in the west," and there is no possibility of sulfuric deposition since most air pollutant emissions in power-receiving regions have exceeded the environmental carrying capacity, while there is a relatively large space for sulfuric deposition in the power-sending regions. In addition, power-receiving regions are economically developed, and the population density and per capita gross domestic product (GDP) are much greater than that in power-sending regions, and the economic loss caused by pollutant emissions is much higher than that in the power-sending regions. The research results indicate that the economic loss arising, per unit of sulfur dioxide emissions in the eastern and central regions of China, is 4.5 times that in western and northern regions. By building large mine-mouth power plants in power-sending regions and UHV cross-regional power transmission, it is predicted that by 2020 the sulfur dioxide emissions in the eastern and central regions of China may fall by 550,000 tons/year, and environmental loss may reduce by RMB 4.5 billion/year.

(2)   Reduce carbon dioxide emissions and cope with climate changes.

By strengthening power grid construction and building a smarter power grid, on the one hand, it can effectively improve the comprehensive energy transportation system, enhance the capabilities of the system for receiving clean energy sources, including hydropower, wind power, and solar power, reduce line loss of the power grid, lower coal consumption for power generation, and increase the economic and operational efficiency of the system; on the other hand, it can enhance energy efficiency by promoting equipment upgrading, raise the proportion of electric energy to terminal energy consumption, and facilitate a large-scale development of the electric car. All of the above benefits can directly or indirectly reduce carbon dioxide emissions. By estimate, by 2020, the power industry can reduce carbon dioxide emissions by about 1.8 billion tons, with a contribution rate of[1] about 18%–20% for achieving China's target of reducing carbon emission intensity per unit of GDP by 40%–45% by 2020.

(3)   Reduce the impacts of the power industry particulate matter (PM) 2.5 pollution.

PM 2.5 pollution and smog control have become increasingly severe problems of public concern, and the Chinese government will take stronger measures against such problems. In the future, China will accelerate the

---

[1]The measurement method: contribution ratio of the smart grid on emission reduction = emission reduction of the smart grid in 2020/China's total emission reduction in 2020, of which China's emission reduction in 2020 = GDP in 2020 × carbon emission intensity in 2005 × (40%–45%), the emission reduction is 9 billion–10 billion tons.

development of clean energy, intensify its efforts of adjustment to fossil energy development, and is likely to adjust existing and additional power-generating energy correspondingly. The Chinese government will energetically develop UHV power transmission and build the "North China, East China, and Central China" synchronous power grid, control the total coal consumption on the overall principles of coal and oil substituted with electricity through long-distance transmission, so as to significantly reduce PM 2.5 pollution in Eastern and Central China. It aims to reduce the impact of the power industry on PM 2.5 pollution in the eastern and central regions of China by[2] 22 percentage points by 2020.

(i) The development of a UHV power grid may help realize China's clean energy development goals and efficient utilization of clean energy, reduce fossil energy consumption and pollutant emissions, and significantly reduce the impact of the power industry on PM 2.5 pollution. The impact of the power industry on PM 2.5 pollution in the eastern and central regions of China is expected to drop by 10 percentage points by 2020, compared with that in 2010.

(ii) The development of the UHV power grid may facilitate the optimization of coal-fired power stations, significantly reduce the coal consumption of the power grid of "North China, East China, and Central China," effectively use the cleaned coal and external electricity, and improve the quality of the local environmental. By placing equal emphasis on coal transportation and power transmission and energetically developing coal power bases, the impact of the power industry on PM 2.5 pollution in the eastern and central regions of China is expected to drop by 12 percentage points by 2020 compared with that in 2010.

(4) Save land for energy delivery and increase the benefits of land use.

The land of a coal transportation corridor is fully used, with the characteristic of exclusiveness, while a power transmission corridor is a "highway for overhead energy delivery," under which the land can be utilized conditionally. To estimate, in the case of delivering the same amount of energy, the area occupied for UHV AC power transmission is merely 1/2–1/4 of that for railway-sea coal transportation, thus UHV power transmission can save a lot of land for energy delivery. Meanwhile, building coal power stations in sending-end regions may make way for the development of higher-value land resources in the eastern and central regions of China, increase the utilization efficiency of China's land resources on the whole, significantly reduce the land occupied for coal gangue in coal-producing areas, and save the land for coal storage sites of coal-fired power plants. According to the

---

[2]The "impact of the power industry on the PM 2.5 pollution" refers to the contribution value to regional PM 2.5 concentration (mcg/m^3) under the comprehensive effects of primary particles and secondary particles from the power industry. In the analytical process, the emission reduction of the power industry rather than other industries is considered.

analysis on scenario 1, by 2020, the eastern and central regions of China are expected to see an additional 150 GW of inward coal power flow. By replacing coal power stations in the eastern and central regions of China with UHV transformer stations or convertor stations, it may save a total landmass of about 4500 ha.

6.3.4 SAFETY BENEFITS

(1) Enhance electric power supply security of the eastern and central regions of China.

As an important measure for promoting diversified energy transportation, accelerating the development of cross-regional power transmission may effectively mitigate the pressure of railway transportation and ensure abundant energy supply. The rational distribution of coal transportation and power transmission may help create a mutual-support situation for effectively mitigating the impacts of major natural disasters (see Fig. 6.12) and reducing losses. By energetically developing UHV power transmission and opening up a new way for delivering energy and electricity to the eastern and central regions of China, it may create a development pattern of coal transportation and power transmission by making their respective advantages complementary to each other, laying a firm foundation for establishing an energy transportation system under new circumstances, and effectively enhancing the capability of energy supply in Eastern and Central China.

(2) Raise power supply reliability and reduce power outage loss.

As the power grid gets smarter, it is possible to achieve smart distribution by relying on smart equipment and advanced technologies. Moreover, various sensors will be able to monitor electric parameters such as voltage and current and the operating conditions of important components, provide power companies and system operators with more accurate information on safety conditions of the power grid, assist power grid operators in taking preventive measures in a timely manner, and increase power supply reliability. The capability of the system against failure will increase significantly. In the case of failure, it can minimize the impacts and losses caused by failure. To estimate, the formation of smart grids may reduce the outage cost of China by about RMB 19 billion.

(3) Give impetus to the development of the electric car, and enhance the safety of energy supply.

The construction of the smart grid provides the large-scale use of a battery-type means of transportation with a platform for optimization control and services. Assuming that China will have 5 million electric cars by 2020, thanks to the introduction of relevant smart grid technologies, and each electric car runs 10,000 km/year, electric cars can save about 3.8 million tons of fuel, making China's external dependence on petroleum drop by about 0.5 percentage points.

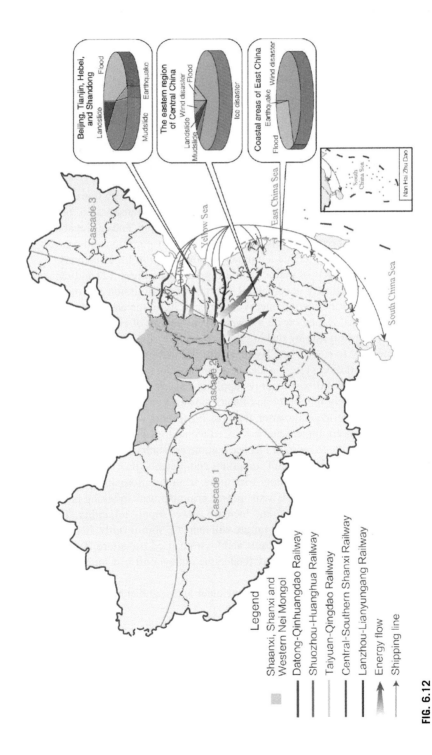

FIG. 6.12

A schematic diagram of capabilities against natural hazards by placing equal emphasis on coal transportation and power transmission.

6.3.5 **SOCIAL BENEFITS**

(1)   Promote the coordinated economic development of China's eastern and western regions.

At present, coal tends to be delivered directly out of coal-producing areas in China. Therefore, the resource advantages of the coal-producing areas cannot be effectively turned into economic advantages, since there is a short industrial chain of coal in coal-producing areas. The construction of a cross-regional power grid may help increase the coal power conversion ratio of the areas where western and northern coal bases are located, extend the industrial chain for coal development and utilization in coal-producing areas, strengthen the effects in driving local economic development, narrow the development gap between different regions of China, optimize regional divisions of work, and promote industrial structural adjustment. Taking Shanxi, for example, the results of calculation and contrast by using the IO model shows that the contribution ratio of coal transportation and power transmission to the GDP growth of Shanxi is about 1:6, and that to employment is about 1:2.

(2)   Promote new product development and the formation of new services market, and drive social and economic development.

The construction of the smart grid can effectively drive the development of relevant traditional industries including electric equipment and device manufacturing industry, the metal-smelting industry, the rolling processing industry, the metal-ware making industry, etc., and promote the development of the hi-tech industries and emerging industries including new energy, new materials, information network, and energy conservation and environmental protection, which are related to the smart grid. According to the dynamic computable general equilibrium (CGE) model, between 2011 and 2020, the smart power grid of State Grid Corporation of China (SGCC) may increase the output of relevant industries of China by about RMB 1.5 trillion, provide 145,000 persons with jobs and raise the labor remuneration of employees by about RMB 3.6 billion yuan/year.

The realization routes to China's non-fossil energy development goal

The route to non-fossil energy development is a guide to action for achieving China's non-fossil energy development goals. On the basis of the analyses on non-fossil energy development scenarios and the supporting role of the power grid, this chapter sets the overall path to non-fossil energy development by stages; in consideration of the development characteristics of various non-fossil energy sources, it gives a detailed analysis on specific development paths and supporting conditions for power generation based on major non-fossil energy sources such as hydropower, nuclear power, wind power, and solar power, and puts forward an implementation plan on achieving the non-fossil energy development goal.

7.1 OVERALL PATH

In accordance with the above analyses, the large-scale utilization of hydropower, nuclear power, and wind power is the key to achieving non-fossil energy development goals. As is seen from the development prospects of power generation based on non-fossil energy sources such as hydropower, nuclear power, and wind power, China has a strong capability of hydropower construction and the problems it faces are mostly resettlement and eco-environmental protection, which can be solved through the joint efforts of competent authorities, power companies, and local governments; the main problems of the nuclear power industry are related to nuclear safety and equipment manufacturing capability. China has no experience in large-scale construction of third-generation nuclear power stations, and its equipment manufacturing capability has uncertainties to some extent; the wind power is mainly restricted by the consumption capability. As is seen from the actual situation, it is hard for China's hydropower, nuclear power, and wind power to reach their respective maximum possible development scale by 2020.

Therefore, to achieve the non-fossil energy development goals, we must adjust the economic structure, raise the efficiency of energy utilization, control the total energy demand, and take a practical and feasible path to speed up the joint development of hydropower, nuclear power, and wind power. Specifically, we shall control the total energy demand, and do our best to keep the total energy demand under 4.8 billion tons of standard coal by 2020, speed up construction of hydropower stations, and strive to attain the goal of an installed hydropower capacity of 350 GW by 2020; we shall develop nuclear power in a safe and efficient way, address China's

Non-Fossil Energy Development in China. https://doi.org/10.1016/B978-0-12-813106-0.00007-6

technological and manufacturing capabilities, actively introduce overseas resources, and strive to achieve the goal of an installed nuclear power capacity of about 60 GW by 2020; we shall actively develop wind power and solar power, coordinate and make proper plans on wind power, solar power, and other power sources and power grid, master core technologies for design, manufacturing, and operation of large-capacity and high-efficiency wind power and solar power units, and meet the requirements on consumption of wind power of more than 250 GW and solar power of more than 100 kW by 2020; we shall promote and utilize other renewable energy sources such as solar energy and biofuel, and ensure that the scale of non-fossil energy for other purposes than power generation exceeds 100 tons of standard coal by 2020.

China's non-fossil energy development path is shown in Fig. 7.1. Specifically, the non-fossil energy development paths for the different development periods of 2015–20 and 2020–30 are shown below.

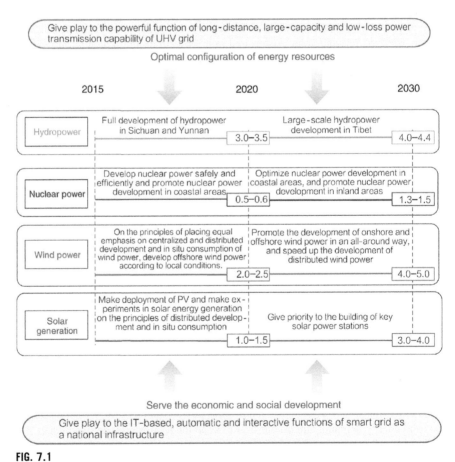

FIG. 7.1

Non-fossil energy development path (unit: 100 GW).

7.1.1 **2015–20**

From 2015 to 2020, the Chinese government intensified its efforts and will continue to strengthen the development and utilization of hydropower, nuclear power, and wind power, build robust smart grids, energetically develop pumped storage power stations, actively develop and demonstrate new energy storage technologies, and strive to increase the cleanliness of new energy supply capabilities.

It is economical and feasible to develop hydropower technologies, which are mature and have large energy density. The main challenges to hydropower development are eco-environmental protection, relocation and resettlement, and overall basin planning. Hydropower is one of the key areas of China's clean energy development in the future, with the highest rate of contribution to the clean energy structure of China before 2030. In order to achieve the target of non-fossil energy use accounting for 15% of the total primary energy consumption by 2020 of the total primary energy consumption, the installed hydropower capacity must reach 300–350 GW, and the rate of hydropower's contribution to China's energy structure for clean energy power generation must exceed 50%. The Chinese government shall continue to promote hydropower construction, particularly the rapid and orderly development of large and medium-sized hydropower stations.

As a major alternative energy, nuclear power features high-energy density and small fuel transportation and has the conditions for technical development. Currently, China has the basis and conditions for large-scale nuclear power development in terms of independent design, independent manufacturing, independent construction, and independent operation, as well as site resource reserves, safety, and economical efficiency of nuclear fuel supply and operation. In the future, the main challenges of large-scale nuclear power development are management systems and mechanisms, nuclear fuel supply, nuclear research levels, as well as site resource planning and protection, etc. Nuclear power has a relatively high rate of contribution for China to develop a clean energy structure. To achieve the target of non-fossil energy use accounting for 15% of the total primary energy consumption of China by 2020, the installed capacity of nuclear power must reach 50–60 GW, and the contribution rate of nuclear power in China's power generation from clean energy shall reach about 30%. Nuclear power shall be deemed as a long-term and key strategy of China and actively promoted.

With basically mature technical and economic conditions, wind power is a renewable energy with the most mature technology and its cost is closest to that of conventional energy, except for hydropower; however, restricted by natural conditions, the utilization time of wind power is lower than that of power generation based on conventional energy, and the annual energy output of installed capacity per kw is relatively low. At present, China's wind power development is still plagued with problems such as inadequate cross-regional power transmission, inconsistency between wind power planning and power grid planning, poor wind power unit design, low technical level of manufacturing, etc. Before 2020, the Chinese government needs to focus on solving the problem of technical feasibility of large-scale wind power utilization, lower the costs to the greatest extent, and work out a complete solution for wind power development, transmission, and consumption. Under the conditions of ensuring technical and economic feasibility, the Chinese government shall intensify its efforts to develop wind power rapidly and in an orderly manner. It is completely possible for China to achieve the goal of reaching an installed wind power capacity of 250 GW by 2020.

China's solar power technology is at the stage of innovation, improvement, and demonstration. Currently, China is working faster to study and overcome the barriers to technical efficiency of large-scale applications of solar energy, and the technology may be considered as a key area of China's energy strategy and its development scale may be gradually raised. Recently, the Chinese government has focused on promoting rooftop grid connection systems in large and medium-sized cities and building large photovoltaic (PV) power stations in western and northern China through carrying out demonstration projects, so as to further enhance relevant technologies and economical efficiency. By intensifying the demonstration of photothermal power generation projects, China aims to create a situation where various solar power technologies complement each other for joint development. It is predicted that China's installed solar power capacity will exceed 100 GW by 2020.

In terms of power grid development, China has become a global leader in ultra-high voltage (UHV) power transmission technology, flexible alternating current (AC) power transmission technology, dispatching automation technology, etc., and has made important progress in flexible grid connection of distributed power generation and new energy power generation. Robust power grids are mainly used for the construction of backbone UHV grids for large-scale cross-regional power transmission, while smart grids are mainly used for solving key technical problems related to smart primary equipment, dispatching technical support systems, flexible power transmission, interactive marketing, distributed energy storage, etc., so as to receive electricity generated from new energy sources such as wind and solar energy to the greatest extent, adapt to users' increasing and diversified demands, and enhance the reliability and overall efficiency of the power grid. China plans to build robust power grids connecting major coal power bases, major hydropower bases, major nuclear power bases, large renewable energy bases, and major load centers by 2020, so as to effectively accelerate the development of various non-fossil energy sources.

Pumped storage is currently an energy storage power source that is mostly likely to be applied on a large scale, and where other energy storage technologies do not have conditions for large-scale application. Among these, such chemical energy storage technologies as flow batteries, lithium ion batteries, sodium-sulfur cells, and lead-acid cells are still at the stage of laboratory demonstration engineering and the initial stage of application demonstration in China. In the future, the appropriate development path of China's energy storage technologies will be as shown below: use pumped storage first where the conditions permit; support the development of flow battery and lithium battery, actively promoting industrialized development and application demonstration projects; guide lead-acid enterprises to intensify technical application and development in the field of energy storage; follow closely the development of sodium-sulfur cells; attach importance to research and development of such technologies as energy storage with super capacitors, compressed air, flywheels, and superconducting magnets and innovation of business models. By 2020, China's large-scale energy storage will mainly rely on pumped storage, the operating capacity of which is expected to reach about 40 GW, not only increasing the safety, stability, economical efficiency, and flexibility of the system, but also promoting large-scale development of clean energy. After 2020, the cost of chemical energy storage is expected to drop rapidly and the large-scale application of chemical energy storage will start, involving flow cells, lithium ion batteries, and sodium-sulfur cells. As is seen from the optional business models for the development of energy storage in China, if there is a sound development mechanism and complete industrial policies, the business models such as power supply side, user side, third-party operator, power grid, and emergency reserve may be applicable to energy storage technology, etc.

7.1.2 **2020–30**

Between 2020 and 2030, new energy supplies are mostly clean energy sources including nuclear energy, water energy, and wind energy. At that time, China will enter the stage of low-carbon growth and shall strive to promote new energy storage technologies.

Before 2030, China shall continue to leverage its mature hydropower technologies and economic advantages, and ensure that the contribution rate of hydropower exceeds 40% in China's clean energy power structure, where the installed hydropower capacity is expected to reach 400–440 GW. The nuclear power technologies will become mature, the contribution rate of nuclear power will account for 40% in the clean energy power generation structure, and the installed nuclear power capacity is expected to reach 130–150 GW. The onshore wind power technologies have matured and offshore wind power technologies will be mature, and for this China shall strive to expand its installed wind power capacity to 400–500 GW. The solar power technologies will be mature and enter the stage of large-scale development. It is predicted that solar power will become a strategic energy around 2030 for its advantages in resources, technology, cost, and environment, with an expected installed capacity of over 300 GW. The Chinese government will continue to develop pumped storage and raise the installed pumped storage capacity to over 100 kW by 2030. The new energy storage technologies will become mature and their costs decline significantly, large-scale applications of such technologies will begin, and electric cars will account for more than 20% of the total cars.

7.2 SPECIFIC PATHS AND SUPPORTING CONDITIONS
7.2.1 HYDROPOWER DEVELOPMENT PATH AND SUPPORTING CONDITIONS
7.2.1.1 The capabilities of hydropower construction

Over years of development, China has nurtured a hydropower design and construction team with a strong capability of large-scale hydropower construction. In the past 60 years, China, by relying on its own strength, built various hydropower stations on large rivers including the Yangtze River and the Yellow River. In the process of its hydropower construction, China has solved many technical problems related to design and construction, made marked progress in hydropower planning, survey, design, construction, and scientific research, made a host of significant achievements, many of which are global leaders. Thanks to the construction of superlarge hydropower stations and dams including the Three Gorges, Xiaowan, Longtan, and Shuibuya, China's hydropower design and construction technologies have gradually occupied a world-leading position.

At present, China is capable of independently researching, designing, manufacturing, and installing various water-turbine power units and meeting the needs for hydropower development. In the 21st century, a host of homemade large-capacity units for the Three Gorges, Longtan, Laxiwa, and Xiaowan hydropower stations were put into operation for power generation, indicating that China has become a world leader in terms of manufacturing, installation, and operation management of large mixed-flow hydropower units.

On the basis of the above analyses, to make China's installed hydropower capacity reach 300–350 GW, an installed hydropower capacity of 10–15 GW shall be launched annually between 2016 and 2020. As is seen from the capabilities of hydropower design and construction, it is possible to achieve the goal by 2020.

7.2.1.2 Planning of hydropower construction

Southwest China is a key area for China's hydropower development. To achieve the goal of an installed hydropower capacity of 300–350 GW by 2020, the Chinese government shall make a plan for the development and preliminary work of hydropower in southwest China, and accelerate the hydropower development there. According to the project progress at an early stage, the arrangements for hydropower development on the lower reach of the Jinsha River, as well as the cascade hydropower stations on the Yalong River and the Dadu River, are shown below.

7.2.1.2.1 Cascade hydropower stations on the lower reach of the Jinsha River

The main stream of the Jinsha River consists of the upper reach, the middle reach, and the lower reach: the upper reach is the section between Zhimenda and Shigu, the middle reach is the section between Shigu and Panzhihua, and the lower reach is the section between Panzhihua and Yibin. The cascade power stations on the upper reach are mainly located in Xizang, and those on the middle reach are located in Yunnan. The cascade hydropower stations on the lower reach of the Jinsha River are shown in Fig. 7.2.

The hydropower stations on the lower reach of the Jinsha River are Wudongde (8.7 GW), Baihetan (14.4 GW), Xi Luodu (13.86 GW), and Xiangjiaba (6.4 GW), with a total installed capacity of 43.36 GW and an average annual energy output of 190 TWh, as is shown in Table 7.1. At present, the Xiangjiaba and Xi Luodu power stations on the lower reach of the Jinsha River have gone into operation and have begun to generate electricity; the construction of Wudongde hydropower station and Baihetan hydropower station has started, and the two hydropower stations are scheduled to be completed and put into operation after 2020.

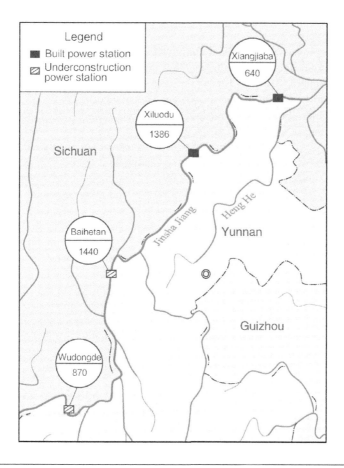

FIG. 7.2

A schematic map of cascade hydropower stations on the lower reach of the Jinsha River.

Table 7.1 Major power source projects of hydropower bases on the lower reach of the Jinsha River

Project name	Planned capacity (10,000 kW)	Status (as of the end 2017)
Jinsha River (lower reach)	4 cascade hydropower stations	
Xi Luodu	1386	Operating
Xiangjiaba	640	Operating
Wudongde	870	Under construction
Baihetan	1440	Under construction
Total	4336	

7.2.1.2.2 Cascade hydropower stations in the Yalong River Basin

There are plans to build 3 reservoirs and 16 hydropower stations on the main stream of the Yalong River.

One reservoir and five cascade hydropower stations are planned to be built on the lower reach of the Yalong River, namely the Jinping stage-I (3.6 GW), Jinping stage-II (4.8 GW), Guandi (2.4 GW), Ertan (3.3 GW), and Tongzilin (0.6 GW), with a total installed capacity of 14.7 GW and an average annual electricity output of 85.5 TWh, of which the Jinping stage-I is the leading power station, and all the hydropower stations have been put into operation.

One reservoir and six cascade hydropower stations are planned to be built on the middle reach of the Yalong River, namely Lianghekou (3 GW), Yagen (1.5 GW), Lenggu (2.72 GW), Mengdigou (1.92 GW), Yangfanggou (1.5 GW), and Kala (1 GW), with a total installed capacity of 11.64 GW and an average annual electricity output of 52.1 TWh, of which the Lianghekou hydropower station is the leading reservoir with a multiyear regulating capacity. Lenggu hydropower station is at the stage of prefeasibility study, and other hydropower stations are under construction and are scheduled to operate during the 13th Five-year Plan period.

The cascade hydropower stations on the upper reach of the Yalong River are at the stage of planning, and are scheduled to go into operation after the 13th Five-year Plan period.

The cascade hydropower stations on the middle and lower reaches of the Yalong River are shown in Fig. 7.3, and the major hydropower source projects on the middle and lower reaches of the river are shown in Table 7.2.

7.2.1.2.3 Cascade hydropower stations in the Dadu River Basin

According to the hydropower development planning of the Dadu River, 2 reservoirs and 23 cascade hydropower stations are now under construction.

It is planned to build one reservoir and nine cascade hydropower stations on the upper reach of the Dadu River, and except for the four hydropower stations that are still at the stage of preliminary planning, there are five hydropower stations, namely

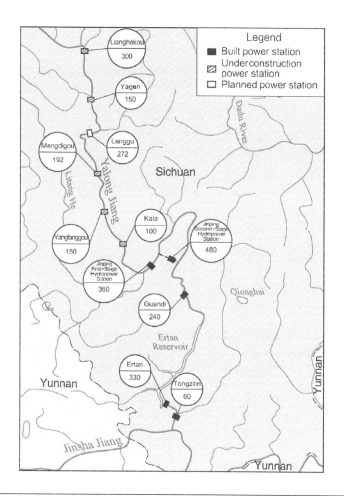

FIG. 7.3

A schematic map of cascade hydropower stations on the middle and lower reaches of the Yalong River.

Shuangjiangkou (2 GW), Jinchuan (0.86 GW), Badi (0.72 GW), Danba (1.54 million kW), and Houziyan (1.7 GW), with a total installed capacity of 6.82 GW and an average annual electricity output of 29.2 TWh. Among these, the leading regulating hydropower station is Shuangjiangkou Hydropower Station that has a multiyear regulating capability. As is seen from the current status of the hydropower stations, except for Shuangjiangkou, Jinchuan, and Houziyan hydropower stations, other hydropower stations are still at the feasibility study or prefeasibility study stage.

There are seven planned hydropower stations on the middle reach of the river, namely Changheba (2.6 GW), Huangjinping (0.81 GW), Luding (0.92 GW), Yingliangbao (1.3 GW), Dagangshan (2.6 GW), Longtoushi (0.7 GW), and Laoyingyan (0.7 GW), with a total installed capacity of 9.63 GW and an average annual electricity

Table 7.2 Major power source projects of cascade hydropower stations on the middle and lower reaches of the Yalong River

Project name	Planned capacity (10,000 kW)	Status (as of the end 2017)
Lower reach	One reservoir and five cascade hydropower stations	
Jinping Stage-I, Jinping Stage-II	840	Operating
Guandi, Tongzilin	300	Operating
Ertan	330	Operating
Middle reach	One reservoir and six cascade hydropower stations	
Lianghekou, Yagen, Mengdigou, Yangfanggou, Kala	892	Under construction
Lenggu	272	Preliminarily approved
Total	2634	

output of 43.8 TWh. As is seen from the progress of the preliminary work, adequate preliminary work has been done for the middle reach of the river, and the construction or relevant feasibility study on all the hydropower stations has started.

There are one planned reservoir and seven planned hydropower stations on the lower reach of the river, of which Pubugou (3.6 GW) hydropower station is a controlling regulating reservoir, and the other six hydropower stations are Shenxigou (0.66 GW), Zhentouba (0.93 GW), Shaping (0.72 GW), Guizui (0.72 GW), Tongjiezi (0.6 GW), and Shawan (0.48 GW), with a total installed capacity of 7.71 GW and an average annual electricity output of 36.8 TWh.

The cascade hydropower stations on the middle and lower reaches of the Dadu River were mostly put into operation during the 11th Five-year Plan and the 12th Five-year Plan periods, and some hydropower stations and cascade hydropower stations on the upper reach were put into operation during the 12th Five-year Plan and the 13th Five-year Plan periods.

Major power source projects of the cascade hydropower stations on the middle and lower reaches of the Dadu River and some cascade hydropower stations on the upper reaches of the river are shown in Fig. 7.4 and Table 7.3.

Besides the planning for the cascade hydropower stations in the above river basins, the installed capacity of planned medium- and small-sized hydropower stations in Sichuan between 2011 and 2020 totals 20 GW, which is mainly concentrated in Western Sichuan. After the above hydropower development plans have been carried out by 2020, the installed hydropower capacity in Sichuan will reach 114 GW, with an increase of about 90 GW between 2011 and 2020. In addition, the installed hydropower capacity of Yunnan is planned to increase by about 50 GW during the "12th Five-year Plan" and the "13th Five-year Plan" periods. As is seen from the

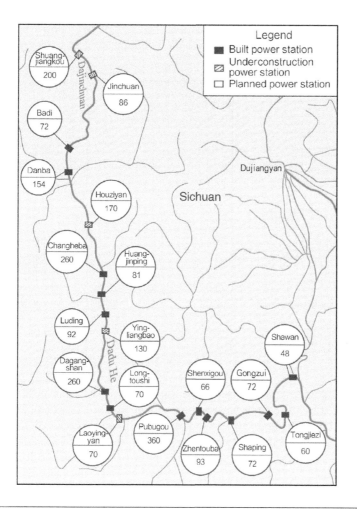

FIG. 7.4

A schematic diagram of cascade hydropower stations on the middle and lower reaches of the Dadu River and some cascade hydropower stations on the upper reach of the river.

development plans on hydropower projects, it is possible to achieve the goal of installed hydropower capacity of 300–350 GW by 2020.

7.2.1.3 Approval of hydropower projects

Since hydropower itself is disputed in terms of environment, relocation, and resettlement, under the influence of the China's polices on hydropower development, the number of approved large- and medium-sized hydropower projects has dropped considerably since 2005. Particularly, the number and scale of hydropower projects approved at the later stage of the "11th Five-year Plan" period are quite limited. Only 16 large- and medium-sized conventional hydropower projects (excluding the

Table 7.3 Major power source projects of the cascade hydropower stations on the middle and lower reaches of the Dadu River and some cascade hydropower stations on the upper reaches of the river

Project name	Planned capacity (10,000 kW)	Status (as of the end 2017)
Lower reach	One reservoir and seven cascade hydropower stations	
Pubugou	360	Operating
Zhentouba, Shaping	165	Operating
Guizui, Tongjiezi, Shenxigou, Shawan	246	Operating
Middle reach	Seven cascade hydropower stations	
Changheba, Dagangshan, Luding	612	Operating
Huangjinping	81	Operating
Yingliangbao, Laoyingyan	200	Under construction
Longtoushi	70	Operating
Upper reach	One reservoir and nine cascade hydropower stations	
Shuangjiangkou, Jinchuan, Houziyan	456	Under construction
Badi, Danba	226	Operating
Total	2416	

underground power source project of the Three Gorges Hydropower Station) have been approved by the central government, with an approved capacity of 9.62 GW, of which the largest hydropower station is the Jishixia hydropower station on the Yellow River, with an installed capacity of 1.02 GW. If the number of hydropower stations approved is still very small during the 12th Five-year Plan period, to achieve the target of installed hydropower capacity of 340 GW by 2020, the Chinese government needs to approve and start more hydropower stations in the later stage of the "12th Five-year Plan" period, which will present a great challenge to China's capabilities of hydropower design, construction, and equipment manufacturing and make it more difficult to achieve the target. It is even difficult to achieve the target of installed capacity of 300 GW, at the end of 2019 or at the beginning of 2020, due to the long period of hydropower construction if the construction progress was slow at the earlier stage.

China has enormous reserves of hydropower projects, and preliminary statistical results show that the installed capacity of hydropower projects meeting the criteria for approval has exceeded 27 GW. A total of 113 hydropower stations are under planning on the Jinsha River (full length), the Yalong River, the Dadu River, the Nu River (full length), the Lancang River (full length), and the main stream of the Yarlung Zangbo River in southwest China, with a total installed capacity of 263 GW. In order to achieve the goal of an installed hydropower capacity of 340 GW by 2020, preparations shall be made on hydropower projects in the early stage and the approval of planned hydropower projects shall be accelerated, so as to ensure that the hydropower projects arranged before the beginning of 2020 can be completed and put into operation.

7.2.2 NUCLEAR POWER DEVELOPMENT PATH AND SUPPORTING CONDITIONS

7.2.2.1 Demands for construction of nuclear power stations

As mentioned before, China's installed nuclear power capacity shall reach 50–60 GW by 2020. It takes 5 or 6 years to build a nuclear power unit. To meet the above target, taking the units that are under construction into consideration, during the period of 2016–20, the average annual output needs to be 4–5 GW, namely four or five 1 GW units. Considering the nonuniformity of nuclear construction, the maximum number of nuclear power stations may reach seven or eight 1 GW units by the 13th Five-year Plan period.

7.2.2.2 Capability of large-scale nuclear power development

The development of nuclear power is an important avenue for China to achieve its target of non-fossil energy accounting for 15% of the total primary energy consumption by 2020. To achieve the installed nuclear power capacity of 50–60 GW by 2020, the Chinese government needs not only to consider the availability of uranium resources, but also the applicability of safe and reliable nuclear power technologies, as well as the possible nuclear power scale and speed.

7.2.2.2.1 Availability of uranium resources

A considerable reserve of natural uranium has been proved in China. By relying on its domestic resources and making a good use of foreign resources, the resources and production capability may meet the demands of China's nuclear power development.

7.2.2.2.2 Applicability of nuclear power technologies

The Chinese government may use existing mature nuclear power technologies to achieve the rapid development of nuclear power. Most of the 400-plus operating nuclear power units in the world use "second-generation" nuclear power technology, which is a proven nuclear technology with high safety, reliability, and market competitiveness. At present, China has mastered the "second-generation" technology that has been constantly improved and can be used on a large scale. The safety and reliability of the technology are higher than that of technologies for most nuclear power stations, which are operating or are about to extend their service life from 40 years to 60 or even 80 years. Even if only existing mature technologies are used, it can support the rapid development of nuclear power in the short or medium term. If the more sophisticated "third-generation" nuclear power technology, which is being introduced, can be proved in actual applications, and its economical efficiency meets the expectations, then China will have more choices for its nuclear power development. It should be noted that, China's State Council convened an executive meeting on October 24, 2012, laying emphasis on raising the admittance threshold, building new nuclear power projects in accordance with the highest safety requirements in the world, and all new nuclear power units must comply with the safety standard of "third-generation" nuclear power technologies.

7.2.2.2.3 The capability of nuclear power construction

China has made a major breakthrough in the development of key nuclear power equipments. China is now capable of manufacturing six to eight units of nuclear power equipment a year, and over 85% of its "third-generation" nuclear power equipment is homemade, compared with 30% in 2008. A host of heavy-duty equipment including "third-generation" nuclear power station pressure vessels, steam generators, and main pipelines are made in China, the model machines of the core equipment including shielded motor main pumps, digital instrument-control systems, and explosion valves have been manufactured, high-temperature reactor control rod drive packages, and fuel loading and unloading systems have been supplied, giving a great push to equipment manufacturers for reaching a new level. China has the capability of construction and installation of at least eight units in at least eight factory sites, has basically mastered the capabilities of nuclear power operation and technical services, established laws, regulations, and systems for nuclear safety management and supervision that are in line with international standards, and mastered the capabilities of carrying out total-process and all-around supervision and management of existing nuclear facilities. In general, China has the conditions for carrying out large-scale construction of "second-generation" nuclear power stations, providing a basic guarantee for achieving the planning goal.

7.2.2.2.4 The speed of nuclear power construction

The countries including the United States and France, which experienced large-scale and rapid nuclear power development, put 10 and 8 units into operation each year, respectively, during the peak period of nuclear power construction in the 1970s and the 1980s. The United States achieved an average annual installed nuclear power capacity of 6.28 GW for 6 consecutive years, hitting a record of 9 GW in a single year. By addressing its late-mover advantages and taking measures such as standard design and one-site-for-more-units continuous construction, China's average annual installed nuclear power capacity may reach 5–6 GW during the later period between 2010 and 2020, and it is possible to achieve the goal of installed nuclear power capacity of 50–60 GW by 2020.

7.2.2.2.5 The building of a talent team

Talents are essential to large-scale and sustained nuclear power development, which has a huge demand for talents. By estimate, a 1 GW nuclear power station needs 400–500 persons, by 2020, about 30,000 persons will be needed for the total installed nuclear power capacity of 60 GW, excluding senior talents for nuclear equipment manufacturing, research, design, and fuel. To build a talent team, we need to start with education, work faster on training, and the implementation of the talent training system. We shall work hard to build a professional and technical talent team, cultivate a host of world-leading senior experts, intensify the training of innovative professionals in colleges and universities, and of skilled workers; we shall employ excellent scientific and technological talents from home and abroad, encourage Chinese students studying abroad to return home or work for China in other ways, and attract and hire high-caliber overseas professionals.

7.2.3 DEVELOPMENT PATH AND SUPPORTING CONDITIONS OF NONWATER RENEWABLE ENERGY SOURCES INCLUDING WIND POWER AND SOLAR POWER

7.2.3.1 Technological level of wind power and solar power

During the 12th Five-year Plan period, China's installed wind power capacity grew rapidly and the development situation kept improving, the technical level increased significantly, nearly all wind power equipment of the industrial chain was made in China, the industrial concentration kept rising, and many enterprises ranked in the top 10 of the world. The technological level and reliability of wind power equipment continued to increase and had basically attained the world-class level. The wind power equipment met the domestic market needs and was exported to 28 countries and regions. The wind power units were more adapted to special environments such as high altitude, low temperature, and frost, and were more easily connected to the grid. Moreover, the economical efficiency of low-speed wind power development technologies was significantly improved, and the quantity of technically feasible wind power resources increased significantly.

In terms of PV development, China's internationalization level of PV manufacturing kept rising, and international competitiveness continued to increase. In 2015, China produced 165,000 tons of polycrystalline silicon, accounting for 48% of global output; its output of PV components reached 46 GW, taking up a share of 70% in the global market. China is capable of manufacturing most key PV equipment and has gradually promoted smart manufacturing, taking a leading position in the world. The PV power generation technology advances rapidly, and the costs keep falling. Chinese firms have mastered the improved 10,000-ton polycrystalline silicon production technology based on the Siemens process, and production based on fluid bed reactor has begun to be industrialized. The average conversion efficiency of monocrystalline silicon cell and polycrystalline silicon cell has reached 19.5% and 18.3%, respectively, taking a leading position in the world. The costs of polycrystalline silicon, PV cells, and components dropped significantly, the cost of PV powers station systems fell to about RMB 7/W, and the cost of PV power generation dropped by over 60% during the 12th Five-year Plan period.

7.2.3.2 Grid connection and consumption of wind power and solar power

Grid connection and consumption are essential to achieving the goal of an installed wind power capacity of 250 GW and an installed solar power capacity of 100 GW. Wind and solar power output is random and intermittent, and in most cases, wind power has the characteristic of antipeak shaving; wind power and solar power have a small guaranteed capacity and can only limitedly replace conventional installed power capacity. The large-scale development of wind power and solar power will not only increase the total installed power capacity and total investment demands of the system, but also increase the peak shaving capacity and spare capacity demands of the system.

To achieve large-scale development of wind power and solar power, we must accelerate the building of the peak shaving capability of the system, actively develop pumped storage power stations, building gas power stations of a certain scale in areas with appropriate conditions, and develop high-parameter and large-capacity thermal power units with good peak shaving capability. It is predicted that by 2020 China's installed capacity of pumped storage power stations will exceed 40 GW, accounting for about 2% of the total power capacity; the installed gas power station capacity will reach about 100 GW, accounting for about 5%.

To achieve large-scale development of wind power and solar power, we must accelerate the development of cross-regional corridors for power transmission. The distribution of wind energy and solar energy resources is not consistent with that of the power load centers in China. As a result of the large-scale and centralized development of wind power and solar power, we need to strengthen cross-provincial and cross-regional power grid interconnection, and expand the scope and scale of consumption. Currently, restricted by such factors as local power demand, power source regulation capability, and power grid development level, the "Three Norths" region has a limited consumption capability of wind power and has to face the prominent problem of wind curtailment. In order to promote further development and utilization of wind power, it is necessary to connect wind power to regional power grids and even the power grids outside the region.

Make the power grid smarter and provide the development of renewable energy sources including wind power with all-around technical support. In the process of power generation, the integrated and advanced information, automation, and energy storage technologies may be used to conduct power prediction and optimal scheduling for renewable energy sources including wind power, support high-proportional grid connection of electricity generated from renewable energy, improve the power output characteristics of clean energy power generation, mitigate the problem of difficult grid-connected operation of wind power stations during the load valley period at night, and solve the problems related to the safe and stable operation of the power grid caused by large-scale clean energy access. In the process of grid connection, we shall work faster to build smart grids, making use of the advanced technologies for power grid operation control and optimal scheduling, effectively enhance the wind power-receiving and power-sending capabilities of the sending-end power grid, ensure the capabilities of the power grid for large-scale, high-efficiency wind power distribution, and optimize the capability for receiving wind power at the receiving-end power grid. In the process of power utilization, we shall make use of the integrated and flexible regulating loads of the smart grid, as well as technologies of cold storage, thermal storage, electric cars, new energy storage, demand side response, etc., actively guide users to use electricity during the valley period, increase the flexibility of power consumption, and promote positive interactions between relevant parties and high-efficient consumption of renewable energy.

Policies and measures for achieving non-fossil energy development goals

8.1 FOLLOW THE STRATEGY OF "GIVING PRIORITY TO ENERGY CONSERVATION" AND REASONABLY CONTROL TOTAL ENERGY CONSUMPTION

To achieve the goal of energy structural optimization and ensure that the non-fossil energy use accounts for 15% of the total primary energy consumption of China by 2020, the Chinese government must strictly control the growth of energy consumption and strive to control the total energy consumption to under 5 billion tons of standard coal by 2020. On the one hand, China's energy demand growth has slowed down, but its total amount of energy still shows a trend of continuous growth. China still has a huge potential for total energy demand growth for some time to come, and its energy resource endowment determines that coal is the most practical available energy; on the other hand, restricted by technical conditions and economical efficiency, there is only limited available clean energy in the short term. Therefore, the Chinese government must attach great importance to energy conservation and control the growth of total energy demand; otherwise, it will be hard to achieve the goal of energy structural adjustment. Effective measures shall be taken to strive to control China's total energy consumption to within 5 billion tons of standard coal by 2020, further reduce the energy consumption growth speed between 2020 and 2030, and work toward controlling the total energy consumption to under RMB 6 billion tons by 2030.

To effectively control the total energy consumption in the future, we must create favorable conditions and make solid progress in the following aspects.

(1) Speed up the transformation of the growth model, channel great energy into making adjustments in the structure of industry. Control total energy consumption from the demand side, effectively reverse the models of inefficient and blind development and open-type energy consumption by making energy constraints serve as a check, make structural improvement and technical innovations and enhance management quality, achieve economic growth by shifting from relying mainly on increasing consumption of energy resources to scientific and technological advances, and management innovations. Improve the system and mechanisms, establish an energy price system that fully reflects the scarcity degree and environment cost of energy resources, and promote efficient and rational allocation of resources; remove

Non-Fossil Energy Development in China. https://doi.org/10.1016/B978-0-12-813106-0.00008-8

market access restrictions, provide society, enterprises and individuals with effective incentives, promote technical innovation and industrial structural optimization and upgrading; promote the development of the service sector and emerging industries, and further increase the proportion of the service sector and emerging industries.

(2) Strictly control growth of industries with high energy consumption, high emissions, and overcapacity, and work faster to shut down outdated production facilities. Put strong constraints on the total quantity of industries with high energy consumption, high emissions and overcapacity such as iron and steel, cement, and electrolytic aluminum, ensure that the total consumption is reduced, and the production capacity of new, renovated, and expanded projects remains unchanged or is reduced; further raise the threshold of industrial access, impose strict requirements on energy saving and environmental protection, require the indicators of energy efficiency and environmental protection to attain a domestically leading level and energy efficiency of existing production capacity to reach the standard within a definite time; work faster to develop implementation plans on shutting down outdated production facilities in key industries during the 13th Five-year Plan period, and break down the tasks to regions by year. Improve the mechanism governing shutting down outdated production facilities, and guide and urge the resettlement of workers laid off due to the shutdown of those facilities.

(3) Excise strict control over total coal consumption, and improve the utilization efficiency of coal resources. Strictly control the burning of untreated coal, lower the proportion of terminal consumption of coal, encourage the replacement of untreated coal, particularly civil-use untreated coal, with natural gas and electricity. Ensure that coal is used in a centralized and efficient way, emphasize increasing the proportion of coal use for power generation in the consumption structure of coal, and raise the level of step utilization of coal resources. Conduct coal reduction control in key areas. Reduce the coal consumption of new fire-burning facilities in Beijing, Tianjin, Hebei and Shandong, the Yangtze River Delta, and the Pearl River Delta, while replacing coal with other energy sources for new fire-burning facilities in other key areas, where the air pollution needs to be controlled. Increase the energy-efficiency level of the major coal-consuming industries, and work harder for energy conservation and emission reduction. Work hard to make the average coal consumption of the thermal power industry to reach a globally leading level at the end of the 13th Five-year Plan period; make greater efforts to shut down outdated production facilities in key iron and steel producing provinces in eastern China, raise the energy-efficiency standard of iron and steel projects that have moved to central and western China; control the gross scale of the coal chemical industry; encourage and guide the building material industry to carry out energy-efficiency benchmarking and assessment work.

(4) Strengthen the management of energy conservation, and establish a long-term mechanism for energy conservation. Establish a mechanism for breakdown and implementation of the target of energy saving. Carry out special action plans on energy saving mainly in the fields of industry and construction. Set targets for energy conservation in key industries and fulfill industrial responsibilities. Make energy-saving renovations in existing buildings, and promote green and energy-saving buildings. Promote energy conservation by using technologies, and energetically increase efficiency of energy utilization. Enhance the technical level of enterprises through technological renovation, development of new technology and IT applications, etc., and promote the applications of new technologies and products through technical standardization. Work actively to establish a policy system for energy conservation, and work out diversified policy-based adjustment measures for laying equal emphasis on intervention and support at the production side, and guidance and supervision at the demand side. Make great efforts to promote contracted energy management, and promote energy saving and consumption reduction through market operations. Lay emphasis on the coordination between government guidance and market regulation, and create an environment in which everyone saves energy.

8.2 WORK FASTER TO MAKE A BREAKTHROUGH IN ADDRESSING OBSTRUCTIONS CAUSED BY CERTAIN SYSTEMS AND MECHANISMS, AND PROMOTE NON-FOSSIL ENERGY DEVELOPMENT

We must change our mindset, work faster to make a breakthrough in addressing obstructions caused by certain systems and mechanisms, and ensure the accelerating development of hydropower, nuclear power, and wind power.

In terms of the system, China's energy development involves not only coal, oil, natural gas, new energy, and electricity, but more importantly, it concerns coordination between different energy sources, as well as mutual coordination between the overall strategy, planning, policy, and technical advances and external cooperation, which require competent government departments to transform their functions. We shall work harder on the top-level design of energy strategy and planning that are important to the overall development of China, improve the investment management mechanism, in which government macro-management policies are coordinated with corporate independent decision-making in an organic way; we shall establish a legal and regulatory system that is related to energy and electric power and adapted to the requirements on market-based reform, and remain coordinated and unified with the development of relevant industries; we shall lower the intensity of energy consumption, raise the effect of the power grid on the comprehensive energy transportation system, strengthen the power grid as a platform for optimization and allocation of energy resources, and raise the position of electric power in China's energy industrial system.

In terms of mechanism, we shall improve that of integrated planning, establish a planning and adjustment mechanism that is led by government and participated in by enterprises, feature transparency in information and democratic decision-making; we shall establish a scientific electricity price system and mechanism, promote optimal resource allocation, optimal system operation and efficient energy utilization, keep enhancing the level of electrification, and establish a sound market mechanism. We shall establish, in due course, a unified, open, competitive, and orderly energy market system that is in line with the national conditions of China, improve the supervisory measures, and gradually establish a supervisory mechanism that is consistent with market-based reform.

8.2.1 **HYDROPOWER**

(1) Intensify policy-based supports, change the thinking on resettlement caused by hydropower construction in an innovative way, and solve the problem related to resettlement for hydropower construction. The Chinese government must attach great importance to resettlement for hydropower, fundamentally address the issue of resettlement, and ensure the implementation of hydropower construction planning. We shall make resettlement an important part of the hydropower project, ensure that measures are taken for planning, design, investment, construction, and acceptance. The key to resettlement is to further establish sound policies and regulations and improve the government's work. We shall adopt a thinking on resettlement in an innovative way, resettle affected residents flexibly in urban areas and urban-rural fringes, for agricultural production, or by means of stock ownership and profit sharing, thus providing them with more choices and truly realizing the goal to "ensure that the affected residents can be resettled, are willing to live where they resettle, and will lead a better life"; we shall establish a reasonable cost accounting mechanism for hydropower development, and include all the expenses for resettlement into the budgetary estimates for hydropower projects; we shall put people first, resettle those who have been affected before the start of hydropower construction, thereby assisting them in getting rid of poverty and becoming better off; we shall publicize the successful experience in carrying out the "Three Gorges" project, and set up a special state-level immigration agency to coordinate and solve relevant issues related to immigration; we shall study how to establish a permanent mechanism by which the rights of land-use and habitation are used for share participation as equity assets, and project benefits can be shared on a long-term basis, and establish a mechanism for reasonable benefit-sharing among immigrants, hydropower companies, the area where the hydropower project is located, and the area benefiting from the hydropower project for the purpose of establishing the socialist market economy, on the principle of integrating responsibilities, rights, and benefits.

(2) Attach greater importance to environmental protection and establish a scientific ecological and environment impact assessment system. We shall always attach great importance to environmental protection in the entire process of hydropower development, strictly putting into practice various environmental protection measures, including planned environmental impact assessment, project environmental impact assessment, as well as environmental supervision and monitoring, etc. We shall improve the relevant systems and laws related to ecological compensation for hydropower development, establish a mechanism for ecological compensation for hydropower development, adjust the distribution relationship of stakeholders in terms of ecological and economic interests, promote ecological and environmental protection, and establish a special fund for ecological compensation for hydropower resource development, on the basis of government compensation and market compensation. The government shall entrust authoritative agencies and organizations to actively conduct basic research on the environmental impact of hydropower development, explore how to establish a set of scientific and mature systems and methods for eco-environmental evaluation, and work actively to promote and strictly implement the technical standards related to hydropower project environment.

(3) Set an annual target of approval of hydropower projects, establish a mechanism for project evaluation, and control the time limit of approval. During the "12th Five-year Plan" period, restricted by such factors as environmental impact evaluation, the number of hydropower stations under construction was far from the set target. Hydropower construction needs a long period, so it is hard to meet the target of an installed capacity of 350 million kW at the end of the 13th Five-year Plan period, on the basis of the progress of current projects. It is recommended to set an annual target of approval of hydropower projects and establish a mechanism for project evaluation so as to ensure that the plans can be implemented. We shall strengthen the coordination and communication between environmental protection, land resources, water conservancy, and earthquake administrations, and do a good job in guiding and supervising matters related to eco-environmental protection, construction land, water resource utilization, water and soil conservation, and earthquakes, etc., and jointly implement relevant plans. For those planned key hydropower projects complying with relevant conditions, the review formalities shall be simple and completed as soon as possible. Particularly, projects including leading hydropower stations, that have a long construction period and a large scale, and which play an important role in cascade development, will be given emphasis and the time of review will be limited so as to start the construction as early as possible.

(4) Accelerate the review process of hydropower planning and attach importance to the development planning of international rivers. In order to support the construction of the major hydropower bases including the Nu River, the Lancang River, and the Yarlung Zangbo River in southwest China, the Chinese government shall work faster on reviewing hydropower planning for the relevant basins and advance the preliminary work related to hydropower, so as to lay a solid foundation for achieving the hydropower development planning goal for 2020 and beyond. At present, China's international rivers are developed and utilized at a very low level, and the main streams of the Nu River and the Yarlung Zangbo River have not yet been developed. It is recommended that the Chinese government make overall plans in accordance with the national energy development strategy and work faster to review hydropower development plans on international rivers.

(5) Strengthen the planning and operation management of hydropower stations, and raise the efficiency of hydropower utilization. On the one hand, we shall make reasonable plans on hydropower for self-use and transmission on a short-term, medium-term, and long-term basis. We shall give comprehensive consideration to such factors as resource conditions, development conditions, and preliminary work when planning hydropower base development, and determine installed hydropower capacity on the basis of electricity consumption growth in power-sending and power-receiving areas. The determination of hydropower transmission plan shall be based on the power source structure of the power-receiving areas, so as to reasonably optimize regional power source structures by hydropower transmission, fully leverage the energy benefits and capacity benefits of hydropower, replace installed coal power capacity in power-receiving areas as much as possible, and ensure efficient utilization of hydropower resources in power-receiving areas. On the other hand, we shall consider the hydropower demands for local use and the growth of local loads at the power-sending ends, ensure complementation, sustained supply, and regulation of such resources as hydropower, thermal power, and wind power on a larger scale (such as hydropower in Xizang and thermal power in northwest China, etc.), and ensure no water curtailment during the flood season and electric power shortage during the dry season.

8.2.2 NUCLEAR POWER

(1) Establish a sound mechanism for nuclear energy development and ensure the sustainability of nuclear power policies. We should have a strong commitment to the idea that nuclear power is the largest scale and most efficient, economical, and safest energy among non-fossil energy sources, never relaxing or abandoning the utilization of nuclear power due to individual incidents or for political reasons. The Chinese government has put forward the principle of "actively develop nuclear power," and formulated the *Mid- and Long-term Development Plan on Nuclear Power*, it is recommended to change the nuclear power development system, based on government directive, and to release basic laws and regulations on nuclear energy including the *Atomic Energy Law*. We shall speed up the establishment of a sound system of laws on nuclear energy, define the status and development prospect of nuclear energy in China's energy system in a legal form, and provide safeguards for the long-term development of nuclear power.

(2) Strengthen the top-level design, set goals on the basis of the national energy development strategy and make overall plans on nuclear development. We shall, at a national level, give overall consideration to the such factors as energy structural transformation, mid- and long-term power supply-demand balance in inland provinces, and smog control in central and eastern China, weigh in on the necessity and potential risks of nuclear power development in the hinterland of China, and make decisions as soon as possible on whether to develop nuclear power in the hinterland of China, as well as the opportunities and route selection for development. We shall make plans for the development scale, distribution, and construction sequence of nuclear power stations in the hinterland of China at the national level.

(3) Unify the technical route, and enhance the technical level and economic competitiveness of China's nuclear power industry. Due to the great difficulty of design and construction of nuclear power stations, complicated product technology, and high safety requirements, the decision makers shall define a unified technical route as soon as possible, develop nuclear power technologies and establish a team of nuclear power professionals. From the perspective of safety, technical maturity and advancement, economical efficiency of various reactors, as well as the supporting capacity of China's nuclear power manufacturing industry, ACPR1000 is a realistic choice for China's mass construction of nuclear power stations during the 13th Five-year Plan period (before 2017). After effectively overcoming the technical barriers, AP1000 shall be adopted as the main type of nuclear station in China in the later stage of the 13th Five-year Plan period; "Hualong No. 1 Nuclear Station" may be used as a complement to China's nuclear stations and a major nuclear unit for export during the 13th Five-year Plan period; after the technical development and demonstration project construction of CAP1400 have achieved the expected effects of major science and technology programs, the power station can be used as a major nuclear power unit for China's economic development and the "go global" strategy on a mid- and long-term basis.

(4) Encourage more parties to make investment. We shall establish a system of nuclear power development in an innovative way, work harder to foster and support investors of nuclear power, grant the enterprises qualified in investment and capable nuclear power station owners, as early as possible, with the status of the controlling nuclear power investor, diversify nuclear power investment and owners, so as to expand the investment and financial channels, enhance the strength of nuclear power construction, and provide the large-scale development of nuclear power with strong financial and comprehensive support.

(5) Strengthen the planning and protection of nuclear power plant sites. Making a good plan on nuclear power plant sites is very important for the large-scale development of nuclear power. Nuclear power plant sites shall be reasonably distributed on the basis of the requirements on China's national economic development, energy demands, and environmental protection, and nuclear power plant sites shall be selected in a standardized way. In the future, the Chinese government shall make overall plans on nuclear power development, keep strengthening site planning and development, and promote nuclear power construction in coastal and inland areas in a methodical manner. In addition, since it takes a very long time from site selection to the preliminary development and construction of nuclear power stations, we should strike a balance between site protection and local economic development planning and construction of nuclear power stations, and effectively protect the sites that have been determined.

(6) Ensure the supply of uranium resources and nuclear fuel. The Chinese government shall increase the capacity of supply of uranium resources and nuclear fuel, and ensure that uranium resources will not become a restriction factor in China's nuclear power development. We shall make greater efforts on the development and research of uranium resource exploration technologies, and strive to make a breakthrough in exploration and development of domestic uranium resources. We shall strengthen cooperation with Canada, Kazakhstan, and Australia in the development of uranium resources, and ensure sufficient supply of overseas uranium resources. We shall further strengthen technological development and production capacity construction of domestic nuclear fuel enterprises, establish a strategic reserve system of uranium resources, and ensure that the nuclear fuel supply can meet the requirements for nuclear power development.

(7) Improve the industrial system for nuclear power research and development, and raise the capability of independent equipment manufacturing. On the basis of the current situation of separation between design and manufacturing of nuclear power equipment in China, we shall actively guide design institutes and manufacturing enterprises to integrate their resources, work faster to establish a nuclear equipment industrial cluster with nuclear power unit design at the core and integrating design, manufacturing, and research and development, foster several complete nuclear power equipment suppliers with research and development, design and manufacturing capabilities, and enhance the capability of supplying complete nuclear power equipment. We shall work actively to jointly tackle key problems, make further innovation on the basis of absorbing advances in overseas science and technology, accelerate efforts in mastering core technologies, ensure that nuclear power equipment is made in China, raise the economical efficiency of nuclear power, and ensure the realization of China's nuclear power development goals.

(8) Strengthen the talent team building for nuclear power development and consolidate the foundation of nuclear power development. The foundation for positive development of the nuclear power industry is the strong capability of scientific research and design of nuclear power stations. The nuclear power industry features close links between upstream and downstream technologies, so we must pool our strength for making a technological breakthrough. We need to work more quickly to foster a contingent of nuclear power scientific researchers and designers, enhance their abilities, and expand the scale of the team, work out effective training plans, policies, and measures for training high-level nuclear power talents, work faster to build a talent team for research, design, construction, operation, and maintenance related to nuclear power, so as to lay a firm foundation for the independent innovation and sustained development of nuclear energy.

8.2.3 **WIND POWER**

(1) Promote rapid wind power development by placing equal emphasis on centralized development and distributed development. During the 13th Five-year Plan period, the Chinese government will continue to build nine large wind power bases at Hebei, eastern Nei Mongol, western Nei Mongol, Jilin, Jiuquan (Gansu), Hami (Xinjiang), the coastal areas of Jiangsu and Shandong, and Heilongjiang, focusing on accelerating the construction of wind power transmission corridors. We shall energetically develop distributed wind power particularly in southern, central, and eastern China, explore ways for integrating with the development of other distributed energy sources, and address their respective advantages. Through the construction of offshore wind power demonstration projects, we shall actively carry out technical research on offshore wind energy development, raise the capability of the manufacturing of offshore wind power equipment, and develop system integration capabilities for offshore wind power equipment, construction, operation, and maintenance.

(2) Accelerate the construction of power transmission corridors of wind power bases, and expand the scope of wind power consumption. We shall make overall plans and coordinate wind power development and power grid construction, and expand the scope of wind power consumption on the basis of the construction of power transmission corridors. This is extremely important to enhance the wind power consumption capability of the electric power system and ensure efficient and reliable wind power operation. During the 13th Five-year Plan period, the wind power consumption of large bases will remain poor, and the contradiction between the sustained and rapid growth of installed wind power capacity and local consumption capability will become more prominent. We shall work faster to review and start cross-regional power transmission projects in the areas that are rich in wind power, and expand the wind power market.

(3) Take a combination of measures to raise regional wind power consumption capability and market competitiveness. We shall establish a sound compensation mechanism for peak shaving, address the in-depth peak shaving capacity of thermal power units, encourage thermal power, nuclear power, and self-owned power plants to take part in peak shaving, appropriately accelerate the construction progress of planned pumped storage power stations, and increase the utilization level of pumped storage power stations; we shall promote the in-depth implementation of the quota system of renewable energy, gradually link the quota system indicators with the total energy consumption indicators, carbon emission reduction indicators, as well as energy conservation and emission reduction assessment indicators; we shall expand the wind power application market, and promote the construction of demonstration projects including wind power heat supply and hydrogen production in northern China; we shall work actively to adjust the on-grid wind power tariff at an appropriate time in accordance with the development situation of wind power and cost changes.

8.3 IMPROVE RELEVANT LAWS AND REGULATIONS AND THE POLICY SUPPORT SYSTEM

To establish a complete incentive policy system covering power generation, grid connection and power utilization, we need to formulate, modify, supplement and improve relevant laws, plans, management practices, policies, and technical standards, so as to promote the development of non-fossil energy.

(1) *Improve relevant laws and regulations.* First, China needs to release as soon as possible the *Energy Law*, a fundamental law, and establish basic principles on energy development and utilization; second, we need to work faster to revise the *Renewable Energy Law*, supplement and improve legal clauses related to development, grid connection, and the utilization of clean energy; third, we need to supplement legal clauses for encouraging consumers to user clean energy; fourth, we need to revise the methods for energy conservation management of key energy users, methods for energy-efficiency label management, and methods for certification management of energy-saving products, etc.

(2) *Strengthen grid connection management.* We shall strengthen research and development of core technologies, further enhance the low-voltage ride through the capability of wind/solar power generation, the active and reactive power regulating capability, and the technical performance of core components, including wind power converter photovoltaic (PV) inverter, etc., so as to adapt to the needs of building power-grid-friendly wind power farms and PV power stations. During the system design period of clean energy access, the technical regulations on clean energy grid connection must be strictly followed, and the technical requirements on active power regulation, reactive power regulation, and low-voltage ride must be clearly defined. During the project implementation period, it is stipulated that new energy power generation equipment must pass tests and certification, and wind power generation farms must pass grid connection tests. During the project operation period, we must strictly follow the relevant operation specifications and processes of grid connection to clean energy, strengthen wind/solar power forecast and prediction management, and ensure safe and stable operation of the electric power system.

(3) *Improve the technical standards.* We shall release industrial standards on wind power farm scheduling and operation and specifications on wind power prediction, revise and improve national standards and industrial standards, including the *Technical Regulations on Grid Connection of PV Power Station*, set technical thresholds related to clean energy power grid connection and safety, and encourage clean energy power-generating enterprises to lower the adverse impacts of clean energy power grid connection on safe and stable system operations through technical advancements, with the addition of some special equipment or equipment upgrading. We shall establish a system for mandatory solar power grid connection certification and tests, and raise the level of solar power technology.

(4) *Establish typical design specifications.* We shall establish typical design specifications for wind power farms and solar power stations (electric), and guide normalized and standardized design. We shall actively promote typical electric design at wind power farms, and achieve normalized design of wind power unit step-up substations, electrical circuits, booster stations, and access systems. We shall establish a sound regulatory system for the design and construction of solar power stations, and raise the level of standardization of construction of solar power stations. We shall promote the applications of new materials, technologies and processes, and build environment-friendly and resource-conserving wind power farms and solar power stations that are in line with China's green energy policies.

(5) *Improve the policy support system.* First, improve the benchmark wind power price policy. According to the current situation of different economic development levels, energy supply and demand, and wind resources in different places, the current four types of benchmark wind power prices are broken down by province. Second, make reasonable pricing policies on the transmission of electricity generated from clean energy, including wind power. On the principle that the power source and the power grid are treated equally, in accordance with the policy on supporting new energy, we shall reasonably evaluate the clean energy power grid connection and compensation price level of power transmission projects in large wind power bases. Third, establish a compensation mechanism for auxiliary power source services. We shall fully tap the potential of peak shaving of the electric power system, and grant reasonable economic compensations to those power sources that take part in in-depth peak shaving, frequency modulation and voltage regulation. Fourth, encourage demand-side management and promote various energy storage technologies in areas that are rich in wind energy resources, facilitate peak load shifting, and increase the wind power consumption capability. Fifth, set reasonable electricity pricing and fiscal policies. We shall adjust the compensation mechanism and sharing mechanism of clean energy power pricing, and channel relevant costs into the sale price of electricity. Sixth,

ensure the sufficient supply of funds for encouraging the development of clean energy sources such as wind power and solar power, etc. We shall establish special funds, incorporate them in the fiscal budgets and increase subsidies for clean energy development.

8.4 STRENGTHEN INTEGRATED PLANNING AND PROMOTE SMART GRID CONSTRUCTION

As the proportion of power-generating energy sources and that of electricity in end-use energy rises, the power industry plays a more important role in the energy supply system, and its social and economic impacts have been further raised. The safety, stability, and economical efficiency of the electric power system represents that of the energy supply system. It is recommended to strengthen integrated planning and promote the construction of smart grid, so as to ensure the realization of the goal of non-fossil energy development.

Implement the integrated planning mechanism of the power industry under the new conditions, and realize coordinated development of power sources and power grids. First, establish a sound system of governmental electric power planning management, establish a planning research system for mutual coordination and cooperation under governmental guidance, adjust and improve the organizational system of the planning work, and address the guiding role in the development of the power industry in the future; second, define the relationship of planning between central and local governments, regulate the planning responsibilities of central and local governments, establish appropriate systems, procedures and regulations for planning research, development, consultation, approval, execution, adjustment, summary, and evaluation feedback, etc. We shall fully reflect the underlying role of planning at a lower level with that at a higher level, and the guiding role of planning at a higher level with that at a lower level. For national planning, annual goals shall be set and local development scales (by province) shall be defined so as to guide local planning. Third, establish a rolling adjustment mechanism for electric planning. We shall, in accordance with legal procedures, organize relevant organizations to conduct rolling studies on a regular basis and make rolling adjustment on electric planning. Fourth, incorporate power generation based on non-fossil energy sources into integrated power development planning, strengthen planning by integrating non-fossil energy and other power sources, non-fossil energy and power grids, local consumption and power transmission of non-fossil energy, sending-ends, and receiving-ends of non-fossil energy. The competent authorities shall coordinate non-fossil energy development plans and the development plan of the power industry, and make overall arrangements on the construction layout and sequence of various power sources and power grids at different levels, so as to maximize the role of various non-fossil energy sources in China's energy structural adjustment.

Expedite the construction of the smart grid and utilize clean energy in an efficient way. We shall boost the construction of the smart grid and raise the electric power transmission and distribution level, so as to meet the requirements for the development of new energy sources including wind energy; we shall enhance the energy utilization level and utilization efficiency, and raise the service level. We shall speed up the construction of the cross-regional power grid, expand the development scale and consumption market of clean energy, promote the development of clean energy, and work to realize optimum resource allocation within a larger range. We shall make the power grid smarter, meet the demands for large-scale clean energy power grid connection and "plug and play" of distributed power sources, and maximize the use of clean energy sources such as nuclear power, hydropower, wind power, and solar power, etc.

In order to enhance the promoting role of the smart grid on clean energy development, it is recommended to fully influence the leading role of government, strengthen organization, coordination and guidance in integrated planning of the power industry, in terms of smart grid planning, standard setting, and innovation in business models, etc., make incentive polices in terms of technological project research and development, experimental and demonstration projects as well as promotion and applications, etc., and make relevant policies on clean energy consumption and economic compensation, etc. Meanwhile, we shall strengthen research and development of key smart grid technologies and equipment, based on the actual situation in China, strive to expand our global presence, grab the share of high-end markets of smart grid technology and equipment, and promote the coordinated and win-win development of the whole industrial chain. We shall address the role of the power grid company as a market actor, and encourage relevant parties to take part in and jointly build smart grids, and establish an effective cross-industry communication platform and cooperation mechanism.

8.5 ACCELERATE INDEPENDENT INNOVATION IN ENERGY SCIENCE AND TECHNOLOGY

In the face of the challenge of the third energy conservation and the third industrial revolution, the energy industry is about to experience a technical industrial revolution. The first second energy conservation mainly focused on resource development and utilization technologies, while the third energy conservation mainly focuses on technical innovations in development and application of new energy, and for this we must attach great importance to collaborative innovation in material, equipment manufacturing, operation management, and business models, etc., and pursue innovation-driven development.

To realize non-fossil energy development goals, it is recommended to energetically advance innovations in energy technology as well as demonstration of research and development of key technologies, speed up the development of new energy and the energy-saving service industry, and foster new economic growth areas. By looking forward into the development trends of energy science and technologies, we shall make overall plans on China's development route of energy science and technology, deploy a host of major forward-looking, strategic and cutting-edge and exploratory technologies, and work hard to increase our capacity for making original innovation. Specifically, we shall work faster in research and development of key core technologies of energy conservation, conventional energy, and new energy, etc., make a breakthrough in strategic, frontier and basic key technologies and equipment, lay a firm foundation for the sustained and healthy development of China's energy industry, and promote enhancement and a leap-forward development of energy science and technologies.

The Chinese government needs to make a breakthrough in tackling key energy-related technical problems at different stages by 2030, so as to significantly enhance China's energy-related technological innovation capability over nearly 20 years of effort, change the current situation of heavy reliance on imported core energy technologies and equipment, safeguard energy security in China, gradually raise the market shares of renewable energy, use fossil energy in an efficient and clean way, and enhance the capability of sustainable energy development. Based on the understanding of the current situation and future demands of China's energy science and technology, and by giving an overall consideration to such factors as resources, contribution rate, environment, technology (independent innovation), feasibility and economical efficiency, key energy technologies that are closely related to non-fossil energy development goals involve the following aspects:

(1) Improvements in energy-saving technology and energy efficiency.

In the short term, the Chinese government shall integrate independent development and technology introduction, energetically promote existing mature and advanced energy-saving technologies in particular main sectors such as industry, construction, and transportation with great potential of energy consumption and energy conservation, solve the technical problems related to economic efficiency, practicality, reliability, and the mass production of energy-saving equipment and products, and achieve large-scale applications and significant effects on energy conservation. In the long term, the Chinese government shall channel great energy into developing new and high energy-saving technologies, make a breakthrough in new energy utilization and conversion technologies as well as efficient energy-saving technologies and products, etc., and fully tap the potential of saving coal, oil and electricity.

(2) Efficient, clean, and safe coal development and utilization.

In terms of coal mining, for the key strategic areas for coal development in western and northern China, the Chinese government shall increase the coal recovery ratio by relying on advanced coal mining technologies, carry out activities of high-precision prospecting of coal resources, and adopt a rational, efficient, and environment-friendly development strategy, avoid the supply risks caused by inadequate reserves of precise surveys of coal resources, for which shaft building is possible, low production technology, poor safety, and insufficient production capacity, and provide technologies for coal bed gas resources with a huge development volume and high economic value.

In terms of coal utilization, the Chinese government shall focus on increasing the efficiency of coal utilization, reducing coal consumption, and solving the problem of the combustion of gasified coal. It shall energetically develop and apply new technologies for high-efficiency and low-pollution coal burning, fully master technologies for the design and manufacturing of 600°C ultra-supercritical units with independent intellectual property rights, develop 700°C ultra-supercritical power generation technologies and make initial achievements, carry out technical research on ultra-supercritical circulating fluidized bed units, in due course, on the basis of demonstrating and promoting the 60 kW supercritical circulating fluidized bed unit; we shall carry out research and development of new-generation coal gasification technologies, and fully master the integrated gasification combined cycle (IGCC) technology, on the basis of integrated coal gasification, and carry out relevant research and development; on the basis of technology introduction and independent development, the Chinese government aims to master the most advanced technologies in the world, for coal power water saving, dust removal, sulfur removal, and de-nitration, continues to intensify research and international cooperation on carbon dioxide capture, utilization, and storage technologies, and will implement comprehensive demonstration projects on the basis of the reduction of greenhouse gas emissions in China.

(3) Natural gas power generation technologies.

Make a breakthrough in technologies of manufacturing of large gas turbines, small gas turbines, micro-gas turbines, and internal combustion engines, and provide equipment for the development of gas power stations and distributed electric power systems. By building gas power stations, the Chinese government shall continue to work faster on the introduction of key technologies, work to increase the independent research and development capabilities, build unit test platforms, make an initial technical breakthrough in gas turbine design and high-temperature component manufacturing technology, carry out orderly and phased-in independent research and development and large-scale promotion and application of high-parameter whole-machine gas turbines at the levels of F, G, and H, and significantly increase their shares in China's power generation sector. The gas-steam combined cycle power station has high efficiency (it is possible to achieve the efficiency goal of 70%–75% before 2030) and its performance of carbon dioxide emission is about 60% lower than that of common coal-fired steam unit, so we shall promote the localization of heavy-duty gas turbines with low calorific value by carrying out pilot projects of integrated coal-gasification combined cycle power generation, encourage distributed gas power generation, promote research and development of high-performance and series micro and small gas turbines, and address the advantages (clean, and energy saving, and lower emission) of gas power generation.

(4) Power grid technology.

In order to meet the demands for large-scale and long-distance optimization and distribution of energy resources and grid connection of large-scale power generation based on renewable energy, China needs to increase the power transmission capacity and efficiency of its electric power system, enhance safety of the electric power system, and make it smarter. It is recommended to further promote the use of technologies such as compact and common-tower multicircuit power lines, large-capacity and high-efficient transformers, flexible alternating current/direct current (AC/DC) power transmission, and integrated 3D design, etc., and raise the power transmission capability and design level. We shall strengthen research on development and project demonstrations of key technologies such as $\pm 1100\,kV\,DC$, flexible DC, AC UHV (ultrahigh voltage) series compensation, controllable high-voltage shunt reactors, and large-capacity breakers, etc. We shall carry out in-depth research and application of disaster prevention and reduction technologies for smart grids and power grids, make power grids smarter and raise its antidisaster ability. We shall work to develop and apply large-scale intermittent power source grid connection technologies, distributed power source, micro-grid technology, and simulation technology, and promote the use of renewable energy. We shall support the development of frontier technologies related to large-capacity energy storage and high-temperature superconductivity, so as to make the relevant technologies available for further development.

(5) Hydropower technology.

Continue to enhance dam construction technologies, and make a breakthrough in key technical fields including the technology of 300 m dam construction under complicated geological conditions and high seismic intensity. Further improve the design and manufacturing level of hydropower equipment, and enhance the capability of providing complete machine and relevant services. We shall carry out research and development of large-capacity, low-water-head, and bulb-type units, carry out technical research on 1-million-kW conventional units, ensure localization of pumped storage units with a per-unit capacity of 350 MW and a water head of more than 500 m, and make substantial progress in environment-friendly hydropower development and research on the development and application of river ecological restoration technologies.

(6) Nuclear power technologies.

Keep optimizing the improved Gen-2 large pressurized water reactor nuclear power technology and increase its safety level through independent research and development. Through the introduction of technology, absorb overseas technologies and by making independent innovation, we shall be capable of manufacturing and designing 1-million-kW "third-generation" nuclear power units independently, and design and manufacture advanced large pressurized water reactors, high-temperature gas cooled reactors, fast neutron reactors, and advanced nuclear fuel circulating systems by relying on important and major national science and technology projects. We shall continue to support the research and development of future nuclear energy technologies including nuclear fusion, so as to lay a foundation for the commercialization of technologies in the future, and gradually establish a system with independent intellectual property rights for development, engineering design, equipment manufacturing, and the safe operation of advanced nuclear energy technologies.

(7) Technologies for new energy power generation, grid-connected power transmission and distribution.

China is rich in nonwater renewable energy resources, of which the conversion, transmission, and large-scale use mainly depend on technical advancements and technical innovations. China needs to make a breakthrough in key technologies and strive for a leap-forward development of new energy technologies. In terms of wind power technology, we shall attach great importance to technical advancements of the wind power industry, strengthen research and development of core technologies for wind power units, ensure the independent development of technology, and achieve industrialization of equipment manufacturing and the advancement of wind turbines through large-scale development; we shall energetically develop new technologies and products that can accelerate the coordinated operation of wind power and the power grid, intensify our efforts to make a technological breakthrough in coordination, optimization, and scheduling of wind power and conventional power sources, fault ride-through (FRT) wind power unit/wind power farms, flexible active/reactive power adjustment, and wind power farm cluster control, etc., and accelerate the construction of wind power farms that are friendly to the power grid. In terms of solar power technology, we shall intensify research and development of core components and key technologies such as polycrystalline silicon material, PV solar modules and inverters, solar power heliostats and vacuum tubes, etc., further enhance the technical performance of solar power low-voltage ride-through capabilities and active/reactive power regulation capabilities, set up a test platform for public solar power technical research experiments, and establish a comprehensive system for solar power equipment test and inspection.

Main conclusions

(1) As is seen from the history and trends of energy development in the world, the global energy structure will be characterized by diversification, cleanliness, and low carbon, and the most important feature is the rapid development of clean energy. To support the development and utilization of clean energy on a larger scale and with a higher proportion, it is necessary to allocate a large number of power sources that can be flexibly adjusted and expand the scale of power grid interconnection, so as to better balance the fluctuations of wind power and solar power, and ensure efficient consumption of electricity. Meanwhile, the application of smart power grid technology and micro-power grid technology has become a new feature of power development.

In the next 15 years, coal will remain a major power-generating energy source in the world, the proportion of natural gas to power-generating energy sources will remain relatively stable, while the proportion of non-fossil energy will rise rapidly. It is predicted that the average annual growth of the power-generating energy consumption will be 2.1% between 2015 and 2030. Before 2020, coal will still be the leading power-generating energy source in the world, accounting for 39%. By 2030, the total shares of nuclear power, hydropower, and electricity generated from other renewable energy sources will exceed that of coal power in the power-generating energy structure and approach 40%, and the share of natural gas in power-generating energy sources will remain relatively stable and stand at about 30%.

In order the support the rapid development of wind power and solar power and ensure safe and steady operation of the power grid, the power source structure must be more flexible, and greater importance will be attached to peak shaving power source development and the gradual application of new-type energy storage technologies. In European countries and the United States, the abundant gas power stations and hydropower stations have high peak shaving performance, and flexible power source is an important guarantee for the rapid development of wind power and the safe and steady operation of the power grid. Between 2008 and 2010, the average wind curtailment ratio of the United States was merely 5%. In order to cope with the situation of the grid-connected operation of wind power and solar power at a higher proportion in the future, Portugal and other countries have put forward plans on the construction of pumped storage power stations, and European countries are exploring how to

Non-Fossil Energy Development in China. https://doi.org/10.1016/B978-0-12-813106-0.00009-X

use Norway's capability of hydropower regulation to balance the fluctuations of the wind power of the North Sea and onshore wind power within the scope of the European power grid. In addition, on a long-term basis, the overall breakthrough in the economical efficiency of the new energy storage unit will lay a foundation for the large-scale application of the unit and create conditions for sustained development of new renewable energy sources including wind energy.

Globally, wind power and photovoltaic (PV) power generation are turning from small-scale distributed development and connection to power distribution networks for local consumption to large-scale and centralized development and connection to power grids for balanced consumption within a larger range. In the future, the world's new energy development will be characterized by integration between centralization and decentralization, with the former increasing overtime. At European onshore wind farms, the distributed development of wind resources carried out was initially developed on a small scale, and the electricity generated was connected to the 10–30 or 110 kV power distribution network. However, considering such constraints as resources and land, the potential of distributed development is limited. In order to achieve the goals of clean energy and carbon emission reduction, Britain and Denmark are speeding up to tap the wind resources of the North Sea by building large offshore wind farms, which will converge through high-voltage or extra-high-voltage (EHV) power grids to the European power grid for consumption. Currently, China and the United States are taking a leading position in the world in the construction of large PV power stations.

To achieve high-efficient use of nonwater renewable energy sources including wind power on a large scale and at a higher proportion, we must expand the scope of power balance for the system, and it is very urgent to build a power grid with a higher voltage level and within a larger scope. Meanwhile, the power grid becomes smarter, and micro-power grids are being applied within a larger range. Existing operation experience and grid connection research results indicate that, cross-regional and even transnational power transmission capacity needs to be further expanded for the consumption of large-scale power generation based on renewable energy and for achieving the renewable energy development goals. Addressing the respective technical and economic advantages of alternating current/direct current (AC/DC) power transmission and establishing a larger-scale synchronous power grid will not raise the safety risks of the power grid, the frequency of occurrence of major power outages of main large synchronous power grids in the world does not necessarily link with the scale of the power grid, and good preliminary planning and control management are key to ensure the safe and stable operation of the power grid. In order to meet the challenges caused by large-scale development of wind power in the next 10–20 years, it is necessary to work hard to ensure safe and stable control of the large power grid, coordinated operation of power source and loads, resource conservation and utilization, enhance resource-use efficiency, meet diverse user demands, and ensure smarter application in the aspects of power generation, power transmission, power transformation, power distribution, power consumption, and power scheduling of the electric power system. Meanwhile, the smart multienergy balanced micro-power grid of distributed PV, wind power, gas power, and energy storage will also develop rapidly.

(2) At present, China's energy structure features an excessively high proportion of fossil energy use, a low proportion of clean energy use, inefficient energy development and utilization, an inadequate energy supply, unreasonable energy allocation, increasing eco-environmental impacts and a poor development situation. In order to achieve sustainable development, we must transform the energy development model, adjust the ideas on energy development, and promote the transformation of energy strategy.

With an energy structure in which coal occupies a leading position, China lags behind developed countries in terms of cleanliness of energy structure and non-fossil energy use. In the primary energy consumption structure of China, the proportion of raw coal is at least 35 percentage points higher than that of developed countries, while the proportions of crude oil, natural gas, and nuclear power are lower than that of developed countries by at least 15 percentage points, at least 10 percentage points, and nearly 10 percentage points, respectively. In the power-generating energy structure, the proportion of China's clean energy power generation is at least 10 percentage points lower than that in developed countries. In the terminal energy consumption structure, the proportion of coal consumption in China is 30 percentage points higher than that in developed countries.

China's coal development is plagued with inefficient utilization and an unreasonable transportation system, causing persistent shortages of coal transportation and power transmission, and acute eco-environmental problems. For a long time, since China gives priority to self-balanced power development, and relies increasingly on its western and northern regions including Shanxi, Shaanxi, Nei Mongol, Ningxia, and Xinjiang in terms of coal supply and production, the coal in western and northern regions is transported to eastern and central China via long distances, on a large scale, and through multiple processes, and the proportion of power transmission is excessively low, causing persistent shortages of coal transportation and power transmission and affecting the steady electric power supply in eastern and central China. The price of power coal keeps rising and the energy supply is continuously short, having a severe impact on economic development. Meanwhile, the Yangtze River Delta, the Pearl River Delta, Beijing, Tianjin, Hebei, and Shandong face increasingly acute problems of air pollution by sulfur dioxide, oxynitride, PM2.5, etc., and there is basically no environmental space for coal power development in eastern and central China.

In the future, the Chinese government shall follow the principles of "safe, economical, clean, and efficient" for energy development and take comprehensive measures for "controlling total energy consumption, making structural adjustments, and optimizing transportation." In terms of controlling total energy consumption, China has an insufficient capability of a sustainable supply of conventional fossil energy that is equivalent to 3.6 billion tons of standard coal. For this, the Chinese government must work to raise the efficiency of energy development, conversion, and utilization, and reasonably control the total energy consumption by enhancing the level of electrification. In terms of making adjustments in energy structure, the Chinese government shall promote diversification and cleanliness of energy use. We shall make greater efforts in the exploration and development of oil and gas resources, so as to maintain oil production and increase gas production. We shall actively develop hydropower under the precondition of doing a good job in ecological protection and the resettlement of affected residents, develop nuclear power efficiently on the basis of ensuring safety, effectively develop wind power, and actively develop other new energy sources. In terms of optimizing the system for integrated energy transportation, the Chinese government shall accelerate the construction of cross-regional power grids and increase the capability of the power grid for the optimization and allocation of energy resources. We shall address the comprehensive economic and technical advantages of ultrahigh voltage (UHV) in large-capacity and long-distance power transmission, enhance the status and functions of the power grid in energy transportation, establish a comprehensive modern energy transportation system featuring "equal emphasis on coal transportation and power transmission," promote the large-scale development of clean energy, and reverse the situation of the persistent shortage of coal power transmission.

(3)   On the whole, China's energy resources are characterized by great abundance, variety, and even distribution. In spite of its rich coal, water, wind, and solar energy resources, China is lacking in high-quality fossil energy resources (petroleum and natural gas). China's energy resources are rich in the west and north, and scarce in the east and south. In the future, the production of new energy sources including water, wind, solar, and coal energy sources will be mainly located in western and north China.

It is a reality that China is rich in coal and poor in oil and natural gas resources, but it has relatively abundant non-fossil energy resources including water, wind, and solar energy, which have a good development potential and prospect. China's technically feasible hydropower energy ranks first in the world, the residual proven recoverable coal reserve ranks third in the world, and the residual proven recoverable reserves of petroleum and natural gas rank 14th and 15th in the world, respectively. As is seen from the resource reserves, conventional energy sources (including coal, petroleum, natural gas, and water energy, of which water energy is a kind of renewable energy which can be used for 100 years) have proven that the total residual recoverable reserve of economical available energy resources is 139.2 billion tons of standard coal, accounting for 10.1% of the world's total.

Generally, China's energy resources are rich in the west and north, and scarce in the east and south, and the energy resources are distributed reversely with load centers. As is seen from water energy resources, there is an installed capacity of technically feasible hydropower resources of 572 million kW in China. Geographically, China's water energy resources are unevenly distributed, rich in the west and scarce in the central and western regions. Southwest China (Sichuan, Chongqing, Yunnan, Guizhou, and Xizang) is the region that has the richest water power resources, with its technically feasible resources accounting for 2/3 of those in China. As is seen from wind energy resources, the potential development capacity of China's wind energy resources (50 m above the ground, level 3 and above, the wind power density is not less than $300 W/m^2$) is about 2.38 billion kW, specifically, 90% of the onshore wind energy resources are in the "Three Norths" regions. The three provinces in northeast China, northern Hebei, Nei Mongol, Gansu, Ningxia, and Xinjiang have the largest wind energy resources in China. Among the nine planned large wind power bases, seven are located in the "Three Norths" region, except for Jiangsu and Shandong. As is seen from coal resources, China has 5.6 trillion tons of coal resources with a depth of less than 2000 m, which are very evenly distributed. Geographically, China has a total of 5.2 trillion tons of coal resources to the north of the line of the Kunlun Mountains, the Qinling Mountains, and the Dabie Mountains, accounting for 93% of the total coal resources of China, 5.1 trillion tons of coal resources to the west of the Greater Khingan Mountains, the Taihang Mountains, and the Xuefeng Mountains, accounting for 92% of the total coal resources of China. By province (autonomous region), Shanxi, Shaanxi, Nei Mongol, Ningxia, and Xinjiang have 4.6 trillion tons of coal resources with a depth of less than 2000 m, accounting for 82% of the total coal resources of China.

In the future, China's energy production will be further weighted toward to western and northern regions, and the scale and distance of energy flow will further increase. As is seen from water energy flow, among the 13 planned large-scale hydropower bases, 8 hydropower bases, including the upper stream of the Yangtze River, the Wu River, and the Nanpan River (Hongshuihe River), the upper reach and the northern main stream of the Yellow River, western Hunan, Minzhegan (Fujian, Zhejiang, and Jiangxi), and northeast China hydropower bases are highly developed, with a development and utilization ratio of nearly 70%, and there is little potential of development. The Jinsha River, the Yalong River, the Dadu River, the Lancang River, and the Nu River in southwest China are less developed, with a development and utilization ratio of about 20%, that is, they have a relatively large development potential and form the main hydropower base of China for the future. The hydropower power flow of the "West-East Power Transmission" project will increase. As is seen from the wind energy flow direction, several 10 million kW wind power bases will be built in Xinjiang, Gansu, western Nei Mongol, eastern Nei Mongol, Jilin, and Hebei in the "Three Norths" region in the next 10 years, and will become the main areas for wind power development in China. Restricted by power grid scale and consumption capability, the wind power generated here needs to be delivered to the load centers in eastern and central China for large-scale wind power development and utilization, and China will see that wind power flow from the "Three Norths" to the load centers in eastern and central China. As is seen from the coal flow direction, as a result of cascade development of coal, in the future China's coal production and distribution will be weighted more toward western and northern regions, and the increased output of coal will mainly be distributed in Nei Mongol, Shanxi, Shaanxi, and Xinjiang. On the one hand, the coal flow will continue to show a tendency of "coal transportation from the west to the east, and from the north to the south," on the other hand, the scale of the "West-East Power Transmission, North-South Power Transmission" project will increase significantly.

China is endowed with rich solar energy resources, and has very large development potential and prospects; its development of nuclear energy and biomass energy is picking up pace, and the development scale mainly depends on the capability of fuel supply and the method of resource utilization on a long-term basis. The increasing scale of oil and gas import is the main feature of mid- and long-term solar development. China's northwest and southwest regions have very high solar radiation. China's solar power development has now picked up pace, construction of large grid-connected solar PV power generation bases has started in Gansu and Qinghai, and rooftop PV systems have been promoted in eastern and central China. In the future, as solar power technology matures and the cost falls, the scale of solar energy utilization will further rise. With a relatively large total demand for nuclear energy development and utilization and broad prospects of technical advancements, China has become a country with the largest nuclear power capacity under construction in the world. China has shifted its principle on nuclear power development from "appropriate development" to "positive development," and will adopt the basic principle of "safe and efficient nuclear power development" in the future. China has relatively abundant biomass energy resources, with a large development potential. China has basically realized a balanced development and utilization of biomass power, liquid fuel, biogas, and briquette fuel, but the utilization level is not high and the technical level needs to be improved. China's conventional petroleum resources have a limited scale, its remaining recoverable petroleum reserves are relatively low, so it is hard for the output of crude oil to rise significantly, and the dependence on imports of petroleum will keep rising. In the future, China will meet its demands for petroleum mainly by intensifying exploration, economical use, energy substitution, and diversification of imports. China's proven conventional natural gas resources are deficient and its gas supply depends on imports. In the future, China needs to intensify its efforts to explore natural gas resources and raise the level of development and utilization of nonconventional natural gas, accelerate the construction of natural gas pipelines, inlet pipelines, and liquefied natural gas facilities. People's livelihood shall be given priority at the time of utilization of natural gas resources, distributed gas multigeneration shall be emphasized for the development of gas power, and large gas power stations shall be appropriately developed.

(4) The results of the multiscenario analysis on energy power development in the next 20 years indicate that, to achieve the non-fossil energy development goals, we must base our efforts on control of total energy consumption, rely on the utilization of power generation based on non-fossil energy sources, focus on accelerating hydropower and nuclear power development, and tackle the critical problems of consumption of nonwater renewable energy sources, such as wind power, solar power, etc.

Controlling total energy consumption is an important precondition for achieving non-fossil energy development goals. The multiscenario analysis results indicate that, if the total energy consumption is controlled to under 4.8 billion tons of standard coal, the target can be achieved by 2020; if the total energy demand grows rapidly to 5.1 billion tons by 2020, it will be difficult to achieve the 15% target under the scenario of verification. Therefore, we must attach great importance to energy saving and control the growth of total energy demand. Meanwhile, we may appropriately adjust the statistical standard for energy to incorporate non-fossil energy in noncommercial energy sources in the statistical scope.

The main way for effective utilization of non-fossil energy is electric power generation. Besides a small portion of non-fossil energy sources being used directly for heat supply, gas supply, and production fuel, most non-fossil energy sources are for terminal utilization through power generation, indicating that the power industry plays a central role in achieving the 15% target. It is estimated that, to achieve the target, the proportion of non-fossil energy use to total primary energy consumption needs to reach 12%–13% for power generation, with a contribution rate of more than 80%.

The key to achieving the 15% target is to accelerate the development of hydropower and nuclear power. By 2020, in China's non-fossil energy mix, hydropower will occupy a dominant position, and take up a proportion of more than 40% under any of the scenarios. If the installed hydropower capacity reaches 350 million kW and the installed nuclear power capacity reaches 60 million kW by 2020, it will have a good economical efficiency to achieve the 15% target, due to the small total installed power capacity and the low cost of electric power supply. If the installed hydropower capacity reaches merely 300 million kW and nuclear power 50 million kW by 2020, to achieve the 15% target of under the scenario of high-energy demand, we must accelerate the development of wind power and solar power, build a higher total installed capacity, and pay higher costs.

The key to accelerating the development of fluctuating power sources including wind power and solar power is to solve the problems of consumption. Wind/solar power feature short construction cycles and is not restricted by project construction cycles. It is mainly restricted by the grid connection and consumption capability of the electric power system. To increase the capability of new energy consumption, we need to fully master the conversion, transmission, and distribution technologies for large-scale new energy development, and solve the problems related to economical efficiency and the efficiency of the electric power system. In order to meet the consumption demands of wind power of at least 2.5 million kW and solar power of at least 100 million kW, first, we need to accelerate the construction of cross-regional corridors for power transmission, and scenario analysis results indicate that the proportion of cross-provincial consumption of wind power needs to exceed 50%; second, we shall strengthen the building of the system peak shaving capability, and gradually increase the proportion of power sources of pumped storage and fuel gas that can be flexibly adjusted, to the total installed power capacity of about 9% by 2020, up about 4% over that in 2010.

(5) Large-scale development and utilization of non-fossil energy requires addressing the advantages of optimization and distribution of energy resources via large power grids. The Chinese government shall increase the development scale and utilization efficiency of non-fossil energy in accordance with the functional positioning of large power grids from the aspect of "safe power transmission, efficient distribution, economical operation, and friendly interactions."

Energy structural adjustment and non-fossil energy development have put forward new requirements for the function of China's power grids. In the future, the construction of China's power grids shall meet the requirements on large-scale and long-distance power transmission, coordinated operation of electricity generated from various energy sources, new energy development and new services of power consumption, and building a national power market. Therefore, we must accelerate the construction of the cross-regional power grid and smart grid.

The power grid plays an important role in promoting the development and utilization of non-fossil energy. By 2020, the installed capacity of China's cross-regional power grid for outward hydropower transmission will reach 63 million kW, accounting for 21% of the total hydropower generation of China; the installed capacity of cross-regional power grids for wind power transmission will reach 40 million kW, accounting for about 20% of the total wind power generation of the country; "North China, East China, and Central China" UHV synchronous power grid will receive nuclear power of about 30 million kW, accounting for about 50% of the total nuclear power generation of the country. Research results indicate that, by 2020, China's cross-regional power grids will be capable of transmitting and allocating electricity, generated from non-fossil energy sources, of about 600 billion kWh, equivalent to 180 million tons of standard coal, with a rate of about a 25% contribution, to the target of non-fossil energy use, accounting for 15% of the total primary energy use of China.

(6) The balanced development of nonwater renewable energy sources including hydropower, nuclear power, wind power, and solar power is a rational and feasible path toward achieving China's non-fossil energy development goals and controlling total energy demand. The Chinese government shall place particular emphasis on different aspects of energy development in different stages. By giving overall consideration to the factors such as resource reserves, development conditions, technical maturity, economical efficiency, etc., the Chinese government shall focus on the construction of large coal power bases and non-fossil energy bases on a short-term basis, and further increase the cleanliness of new energy supply on a long-term basis.

The key to achieving non-fossil energy development goals is to control total energy demand and promote the large-scale utilization of hydropower, nuclear power, and wind power. The Chinese government shall take a combination of measures such as establishing systems, setting standards and using strict assessment methods to keep the total energy demand to under 4.8 billion tons of standard coal by 2020, and work faster to ensure that the installed hydropower capacity reaches 300–350 million kW by 2020; we shall actively develop nuclear power and strive to achieve the goal of an installed nuclear power capacity of about 60 million kW by 2020, energetically develop wind power and solar power, make coordinated plans on wind power, solar power, and other power sources and power grids, realize a grid connection and efficient consumption of wind power and solar power of at least 250 and 100 million kW, respectively, promote the use of other renewable energy sources and make it exceed 100 million tons of standard coal.

As is seen from the development path in different stages:
From 2015 to 2020, the Chinese government intensified its efforts and will continue to strengthen the development and utilization of hydropower, nuclear power, and wind power, enhance the efficiency of energy utilization, build smart grids, energetically develop pumped storage power stations, actively develop and demonstrate new energy storage technologies, and strive to increase the cleanliness of new energy supply capabilities.
From 2020 to 2030, the proportion of non-fossil energy sources including nuclear energy, water energy, wind energy, and solar energy to new energy sources will further rise, and China will enter the stage of low-carbon growth, new energy storage technologies be commercialized overtime, and the market share of electric cars will significantly rise.

(7) We shall keep the total energy consumption within the predefined scope, starting with adjusting and optimizing the industrial structure, strengthening the scientific management of energy conservation, strengthening supervision and assessment of energy conservation, reasonably guiding energy consumption, etc. Meanwhile, we shall set up a coordinated and all-around support system for clean energy development from the aspect of system, mechanism, policy, standard, regulation, technology, etc., and facilitate the sustained and sound development of non-fossil energy.

Control the total energy consumption. To effectively control the total energy consumption, we must take comprehensive measures, create favorable conditions, and effectively carry out relevant work in the future: (1) change the economic growth model, adjust and optimize the industrial structure, and achieve economic growth by shifting from relying mainly on increasing consumption of energy resources to scientific and technological advances and management innovations. (2) Strengthen energy conservation management of key energy users, and attach importance to measures for energy conservation taken in such sectors as industry, construction, and transportation. (3) Establish a mechanism for breakdown and implement the target of energy saving, and strengthen supervision and assessment of energy conservation. (4) Enhance guidance of energy consumption, and actively promote the development of energy markets.

Strengthen innovation in systems and mechanisms. We must accelerate the development of hydropower, nuclear power, and wind power. Specifically, we shall focus on striking a balance between relocation and resettlement and environmental protection, work faster on reviewing hydropower projects, make sound policies on nuclear energy development, attach great importance to nuclear power security, encourage the involvement of diversified investors, develop ways for innovative development and utilization of wind power, establish a development mechanism for achieving an all-win situation of developers, power grid operators, users, etc., ensure orderly and healthy development, raise the capability of innovations in wind power technologies, and strengthen construction and management of wind power farms.

Improve the policy support system for non-fossil energy development. We shall make, formulate, revise, complement, and improve relevant laws, policies, and plans on, and management measures and technical standards for non-fossil energy development, strengthen grid connection management of random and intermittent power generation, based on non-fossil energy sources, and form a complete incentive policy system of non-fossil energy covering power generation, grid connection, power transmission, and power consumption.

Make integrated planning and energy technology innovation. We shall address the integrated planning function of the power industry and realize the coordinated development of power sources and power grids, incorporate the development of power generation, based on non-fossil energy sources, on electric power planning, so as to achieve orderly development as planned. We shall accelerate the construction of smart grids, use non-fossil energy efficiently, give great impetus to energy technology innovation as well as development and demonstration of major technologies, work faster on the development of the new energy and energy-saving service industry, and foster new economic growth areas.

Appendix

APPENDIX I. EXPLANATIONS ON THE MAIN STATISTICAL INDICATORS

(I) EXPLANATIONS ON RELEVANT ENERGY INDICATORS OF THE NATIONAL BUREAU OF STATISTICS OF CHINA

Primary energy: a form of energy and resources found in nature that has not been subjected to any human-engineered conversion or transformation process. It consists of raw coal, crude oil, natural gas, oil shale, nuclear energy, solar energy, water energy, wind energy, wave energy, tidal energy, geothermal energy, biomass energy, ocean thermal energy, etc.

Secondary energy: an energy product by conversion from a primary energy, such as electricity, steam, coal gas, gasoline, diesel, heavy oil, liquefied petroleum gas, ethyl alcohol, methane, hydrogen, coke, etc.

Primary energy can be further divided into renewable energy and nonrenewable energy. In nonrenewable energy sources, fossil energy that includes coal, petroleum, and natural gas is currently the main energy source. Other primary energy sources beyond fossil energy are collectively known as non-fossil energy.

Renewable energy: this includes solar energy, water energy, wind energy, biomass energy, wave energy, tidal energy, ocean thermal energy, etc., which can be renewable in nature.

Nonrenewable energy: this includes raw coal, crude oil, natural gas, oil shale, nuclear energy, etc., and is nonrenewable.

Fossil energy: hydrocarbon or its derivatives. As an energy source originating in prehistoric times through sedimentation, it mainly consists of coal, petroleum, and natural gas, and generates heat through the oxidation reaction of hydrocarbon. The emission of carbon dioxide generated from oxidation reaction is one of the major factors of global climate changes, and the emission of carbon dioxide related to fossil energy has become one of the biggest concerns to the population.

Non-fossil energy: other primary energy sources beyond hydrocarbon. It usually has a relatively low or near-zero emission of carbon dioxide, which is a kind of greenhouse gas. For this, it is an environmentally friendly energy form from the perspective of curbing global warming.

Clean energy: energy that generates no hazardous substances in the process of its production and use. It is renewable and can be recovered after being consumed, or is nonrenewable (such as wind energy, water energy, and natural gas), or has been treated by using clean technology (such as clean coal, clean oil, etc.). It is a frequently used but nonstandard concept, and its meaning may vary on different occasions. Generally, all renewable energy and energy sources, that have lower emissions of pollutants and greenhouse gases than coal, are called clean energy in a broad sense.

Total energy production: the total amount of primary energy production of a country (region) within a certain period. It is a total amount indicator for the level, scale, process, composition, and growth rate of national (regional) energy production. Primary energy production includes electric energy output of raw coal, crude oil, natural gas, hydropower, nuclear energy, and other power sources (such as wind energy and geothermal energy), but does not include low-calorific value fuel production, the utilization of biomass energy and solar energy, as well as the production of secondary energy converted from primary energy.

Total energy consumption: the total amount of several energy sources consumed by various industries and residents of a country (region) within a certain period. It is a total amount indicator for the level, composition, and growth rate of energy consumption. Total energy consumption covers raw coal and crude oil and their products, natural gas, and electricity, but does not cover the use of low-calorific value fuel, biomass energy, solar energy, etc. Total energy consumption consists of terminal energy consumption, loss from energy processing and conversion, and loss.

Final energy consumption: the amount of energy consumed by various industries and residents of a country (region) within a certain period, deducting the loss and consumption of secondary energy for processing and conversion.

Loss from energy processing and conversion: the difference between the total amount of a variety of energy products and that of various sources consumed for processing and conversion in a country (region) within a certain period. It is an indicator for measuring the change of energy loss in energy processing and conversion.

Energy loss: the loss caused during the process of energy delivery, distribution, and storage, as well as various losses due to objective reasons within a certain period, excluding the amount of venting and diffusion of various gases.

For making statistics about total primary energy production and total energy consumption, there are, in China, computational results from coal equivalent calculation and electrothermal equivalent calculation in terms of power conversion. For making statistics about the indicators such as the growth rate of energy consumption and energy consumption per CNY (Chinese Yuan) 10,000 yuan *gross domestic product* (GDP), the coal equivalent calculation method is generally used, which, in nature, treats primary power generation from various non-fossil energy sources as coal power equivalent, and does not consider the losses arising from energy processing and conversion in the process of converting fossil fuel into electricity. The concept of total primary energy in this book also complies with practice, namely the amount of energy, corresponding to electricity, is converted on the basis of the coal equivalent calculation method.[1]

---

[1] It should be noted that, there are methods other than the coal equivalent calculation and the electrothermal equivalent at home and abroad for conversion of power generation based on non-fossil energy. Besides the coal equivalent calculation method (also called substitution energy method) and the electrothermal equivalent (also called direct equivalent method), the method adopted by the International Energy Agency (IEA) combines the above two methods. The method converts nuclear power and geothermal energy into primary energy, with an efficiency of 33% and 10%, respectively, and converts other energy sources into primary energy with an efficiency of 100%, which is equivalent to that of the electrothermal equivalent calculation method. There is a certain incomparability between different methods due to the use of different statistical calibers.

Coal equivalent calculation method: the electricity is converted into standard coal in accordance with average coal consumption for thermal power generation. China's coal consumption for thermal power generation was 319 g/(kWh) in 2009, and is expected to be about 305 g/(kWh) by 2020 according to the relevant research report of the National Energy Administration.

Electrothermal equivalent method: this refers to converting electricity into standard coal on the basis of its heat equivalent of work, with a conversion factor of 1 kWh = 0.1229 kg of standard coal.

(II) DEFINITION OF THE RESEARCH SCOPE OF THIS BOOK

According to the standard of classification of the National Bureau of Statistics of China, the non-fossil energy studied in this book covers water energy, nuclear energy, wind energy, solar energy, and biomass energy, as well as other energy sources such as geothermal energy, ocean energy, etc., but does not cover the traditional utilization of straw and firewood. Fossil energy mainly refers to the three major conventional energy sources such as coal, petroleum, and natural gas.

Non-fossil energy is mainly converted for power utilization, and all non-fossil energy sources that are converted into electricity are considered to be commercial energy.

The utilization of non-fossil energy sources for purposes other than power generation mainly covers solar heat utilization, biological gas supply, geothermal heat utilization, solid particle, and biofuel, most of which are noncommercial energy sources. At present, China's statistics about total primary energy consumption do not cover noncommercial energy sources.

According to the *Scientific Development 2030—Research Report on China's National Energy Strategy* of the National Energy Administration, the total non-fossil energy for purposes other than power generation is expected to reach about 130 million tons of standard coal (including noncommercial energy sources, solar heat utilization, biological gas supply, etc., but does not cover traditional the utilization of straw and firewood) by 2020; if the applications of noncommercial non-fossil energy sources such as solar heat and biological gas supply are not considered, it is predicted that the total non-fossil energy for purposes other than power generation will be about 40 million tons of standard coal by 2020.

In this book, full-caliber statistics (including noncommercial and non-fossil energy) are adopted for total non-fossil energy for purposes other than power generation, that is, it is assumed that there is given condition of 130 million tons of standard coal for the research in this book.

In addition, the definitions of power generation based on new energy and clean energy are shown in the following.

New energy: also known as nonconventional energy, various forms of energy beyond traditional energy. It refers to the energy for which the development and utilization are at the beginning, being actively studied, or to be promoted, such as solar energy, geothermal energy, wind energy, ocean energy, biomass energy, nuclear fusion energy, etc. The definition of new energy made at the "United Nations Conference on New and Renewable Sources of Energy" held in 1980 is: on the basis of new technology and new materials, achieves modern development and utilization of traditional renewable energy sources, replacing resource-limited and environmentally polluting fossil energy with inexhaustible renewable energy, and focuses on the development of solar energy, wind energy, biomass energy, tidal energy, geothermal energy, hydrogen energy, and nuclear energy.

Power generation based on clean energy: power generation based on non-fossil energy sources including hydropower, nuclear power, wind power, solar power, and biomass power.

APPENDIX II. TECHNO-ECONOMIC PARAMETERS FOR POWER SOURCE OPTIMIZATION

(I) UNIT RESERVE AND OVERHAUL SCHEDULE

Spinning reserve: as per the maximum load of 8% for each region.

Cold standby: as per the maximum load of 5% for each region.

Time of unit overhaul: nuclear power (50 days/year), coal power (1 million kW, 60 days/year), coal power (600,000 kW, 50 days/year), existing coal power (45 days/year), thermal power (45 days/year), gas power station (20 days/year), hydropower (30 days/year), and pumped power station (30 days/year).

In addition, in the case of the mutual supply of electricity between provinces (autonomous regions and municipalities), the provinces (autonomous regions and municipalities) undertake service capacity mutually, and only undertake spare capacity of their own, that is, cross-provincial (autonomous region and municipality) spare capacity is not considered.

(II) INVESTMENTS IN POWER PLANT, FUEL CONSUMPTION RATE FOR POWER GENERATION AND RELEVANT TECHNO-ECONOMIC PARAMETERS

The *Reference Cost Index Quota for Thermal Power Engineering Quota Design (2011)* shall be referred to for making an investment in thermal power stations; for the fuel consumption rate of existing coal-fired power plants, the average coal consumption of local thermal power plants in 2011 shall be referred to; for the fuel consumption rate of a new power plant, the typical coal consumption of the unit used shall be referred to. The main techno-economic parameters of a coal power unit are shown in Annexed Table II.1.

Typical values shall be used as techno-economic parameters of the nuclear power unit and the gas power unit, see Annexed Table II.2.

There are two types of pumped storage power stations. The techno-economic parameters of fixed-production pumped storage projects are set in accordance with

Annexed Table II.1 Main techno-economic parameters of coal power unit

Type	Existing coal power station	Single-unit 300 MW thermal power	Single-unit 600 MW unit (made in China)
Time of unit overhaul (day)	45	45	50
Unit investment in alternative power source (yuan/kW)		4400	3600
Proportion of fixed operating expenses to investment (%)	3	3	3
Economic life of unit (year)	25	25	25
Minimum technical output rate of unit	0.4–0.9	0.8	0.4
Average coal consumption rate of unit [standard coal (in g)/(kWh)]	330	310	290

Note: *The air-cooled unit is used for installed coal power capacity in the power grid in northwest China, and the average coal consumption rate of 300, 600, and 1000 MW units increases by 20 g/(kWh), 15 g/(kWh), and 15 g/(kWh), respectively.*

Annexed Table II.2 Main techno-economic parameters of nuclear power and gas units

Type	Gas unit	Nuclear power unit
Unit capacity (10,000 kW)	30	100
Time of unit overhaul (day)	20	50
Unit investment in alternative power source (yuan/kW)	3500	12,000
Proportion of fixed operating expenses to investment (%)	4	2
Economic life of unit (year)	25	40
Minimum technical output rate of unit	0	1
Average gas loss of unit [standard m^3/(MWH)]	212	

the budget estimate of project, and those of optional pumped storage power stations for GESP-III optimization are set as the same typical values, as is shown in Annexed Table II.3.

(III) MAIN POWER-GENERATING FUEL PRICES OF DIFFERENT REGIONS

The coal price shall be based on the average delivery price of raw coal in 2010, and the gas price shall be based on the market price of gas in different regions, see Annexed Table II.4.

Annexed Table II.3 Main techno-economic parameters of pumped storage power station

Name	Fixed-operation pumped storage power station	Optional pumped storage power station for optimization
Unit capacity (10,000 kW)	Planned value of project	30
Time of unit overhaul (day)	30	30
Unit investment in alternative power source (yuan/kW)	Budget value of project, about 3500–5000	3500
Proportion of fixed operating expenses to investment (%)	2	2
Economic life of unit (year)	25	25
Conversion efficiency of pumped storage	0.75	0.75

Annexed Table II.4 Prices of main power-generating fuels in different regions

Name	Average price of power-generating fuel				
	North China (including western Nei Mongol)	East China	Central China	Northeast China	Northwest China
Price of coal (price of raw coal, yuan/ton)	500	870	790	460	350
Price of natural gas (yuan/m^3)	2.50	3.00	3.00	3.00	2.00
Price of fuel for nuclear power generation (converted into price of standard coal)	160	160	160	160	160

(IV) THE CAPABILITIES OF DIFFERENT PROVINCES (AUTONOMOUS REGIONS AND MUNICIPALITIES) FOR MAKING INVESTMENTS IN AND POWER TRANSMISSION OF TIE LINES

The current capability of power transmission between provinces (autonomous regions and municipalities) within a regional power grid shall be defined in accordance with the capacity of existing tie lines; the scale of new tie lines shall be based on the *Plan of State Grid Corporation of China on Power Grid Development for the "12th Five-year Plan" Period*; the power grid investments shall be based on the *Control Indicators for Quota Design of Power Grid Projects (2010)*, and the fixed costs for power grid operation shall be 2% of the unit investment.

APPENDIX III. AN INTRODUCTION OF THE MULTIREGIONAL POWER SOURCE EXPANSION AND OPTIMIZATION MODEL

(I) AN OVERVIEW OF THE SOFTWARE

As one of advanced theoretical methods for studying power source expansion of the electric power system, the power source expansion and optimization model of the electric power system aims to define the most economical power source expansion process of the electric power system. Under the precondition of ensuring balanced system loads and power energy volume, the model gives an overall consideration to the distribution, scale, and development plans of various power sources, studies the economical and rational power source structure of the electric power system, reflects the status and market characteristics of electric power resources, analyzes the total system costs and the cost of the selected power sources, breaks down the total system cost, and sorts the candidate power sources in the system on the basis of economical efficiency, so as to fully leverage the resource advantages of various power sources of the system, realize overall operation of the electric power system and power sources under the optimal condition, and achieve the best cost efficiency of construction and operation.

Developed by the State Grid Energy Research Institute (formerly known as the State Power Economic Research Institute), the "multiregional power source expansion and optimization model (GESP-IV)" is used during this research stage for optimal computation. It had been used for the national economic evaluation on the Three Gorges Project, the planning research on the power grid in South China, as well as economic analyses on electric power projects by using loans from the World Bank and the Asian Development Bank.

GESP-IV is a multiregional power source expansion and optimization model for minimizing the total discount cost within the calculation period. The research on minimum cost has a particularly important meaning to the development of power sources. All electric power projects in the electric power system are interconnected, so a new electric power project is likely to change the operating mode of all other projects. Therefore, only in research with minimum expense, that fully considers the relationship between electric power projects, can the economical efficiency be accurately demonstrated. When an evaluated electric power project is included in the minimum expense plan of the system, it indicates that the project is a part of the optimal development plan on the electric power system, and any other replacing plans will raise the cost for the electric power system.

(II) MAIN FUNCTIONS

As the latest edition of the software, GESP-IV further enhances the functions of environmental protection planning to cover limitations on environmental emissions and optimization of pollutant control equipment. Its main functions are shown as follows:

(1) determine the optimal plan on power source expansion of the electric power system within the planning period, so as to ensure optimal installation progress

of various candidate power sources and the rational distribution of the power source system;

(2) define reasonable proportions of various power sources in the planned system;

(3) define the optimal development sequence of power sources including hydropower station;

(4) define reasonable installed capacity of various power source projects;

(5) analysis on environment policies on environmental protection charges and total quantity control, etc.;

(6) evaluate the costs and benefits of pollutant emission control equipment of the electric power system;

(7) choose pollutant emission control equipment of thermal power plants of the electric power system and optimize the installation progress;

(8) analyze the marginal control cost of pollutant emissions of the electric power system; and

(9) make optimal fuel choice of the electric power system.

(III) SYSTEM MODEL

Mixed integer programming is adopted in the GESP-IV model to solve the problem of mid- and long-term optimal power source expansion planning of the power generation system. The overall mathematical description of the model is shown as follows:

(1) General mathematical description of mixed integer programming
The target function:

$$\min F = \mathbf{A}X + \mathbf{B}Y$$

wherein
F—target function
$\mathbf{A}, \mathbf{B}$—constant matrix
X—continuous variable
Y—0–1 variable
The constraints:

$$\mathbf{C}X \geq \mathbf{M}$$
$$\mathbf{D}X - \mathbf{E}Y \leq 0$$
$$Y \leq 1$$
$$X \geq 0$$

wherein
$\mathbf{C}, \mathbf{D}, \mathbf{E}$—constant matrix
$\mathbf{M}$—constant vector

(2) Target function of model
The target function of the model:

$$\min Z = I - S + F + V + E$$

wherein

Z—target function

I—the present value of the total investment costs within the planning period

S—the present value of the final residual value of new fixed assets within the planning period

F—the present value of fixed operation expenses of the system within the planning period

V—the present value of the variable system operation expense within the planning period

E—the present value of the loss caused by lack of power supply within the planning period

(3) Constraints of the model

Certain constraints must be satisfied in the optimization process, and main constraints of the model are shown as follows:

(i) The effective system capacity must be the maximum load plus the required reserves.

(ii) The total output of any power plant shall meet the load demands.

(iii) The output of each power plant shall meet the limit of minimum technical output at any time.

(iv) The output of a power plant shall not be greater than the maximum installed capacity of the power plant.

(v) The generated energy of a hydropower plant shall not exceed its limit.

(vi) The pumped storage power station must meet its constraints on storage capacity as well as the relationship between its generating capacity and pumping power.

(vii) The installed capacity of each type of power plant shall not exceed its limits.

(4) Main output of the model

The output of the model mainly includes the plan on power source expansion at the minimum cost, optimal mode of system operation, total present value of investment (excluding final balance of fixed assets), and operating expenses. A part of the output information is shown as follows:

(i) load curve;

(ii) plan on power source expansion at the minimum cost;

(iii) the yearly investment process of the system and each power plant;

(iv) the optimal operation mode of all power plants;

(v) interval power flow exchange;

(vi) fixed and variable operating costs;

(vii) fuel consumption; and

(viii) other technical and economic indicators of the system and other power plants, etc.

See Attached Fig. III.1 for the principle structure of GESP-IV.

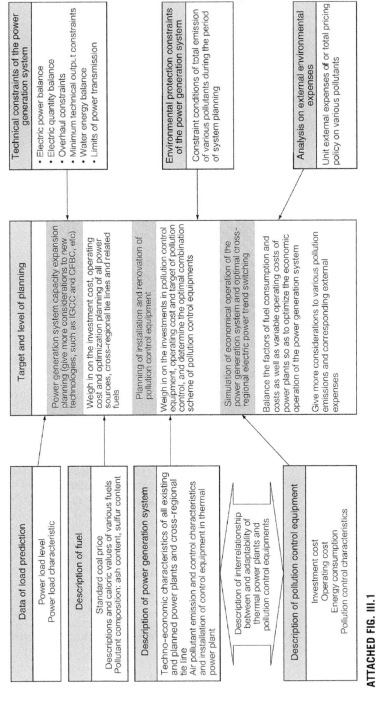

ATTACHED FIG. III.1

The principle structure of GESP-IV.

APPENDIX IV. THE SIMULATION MODEL OF RANDOM PRODUCTION OF THE ELECTRIC POWER SYSTEM

Analysis target: to economically and efficiently meet the demands for operation of the electric power system and consumption of clean energy sources including wind power.

Main constraints: regional 24-h electric power balance on a typical day; maximum transmission capacity of tie lines; regional reserves for load and accident; regional monthly electric quantity balance; regional security startup; regional operation of power stations; unit overhaul, etc.

Output results: results of regional hourly production simulation of the electric power system, including system fuel consumption, operation expenses, emissions of environmental pollutants and greenhouse gases, peak shaving surplus of regional systems on a typical day, cross-regional power exchange, unit operating position and output, and system water/wind curtailment scale and the period of occurrence, etc.

The principle structure of the random production/simulation model of the electric power system, the unit overhaul plan, structure, and process as well as the random production simulation and structure process are shown in Attached Figs. IV.1–IV.3.

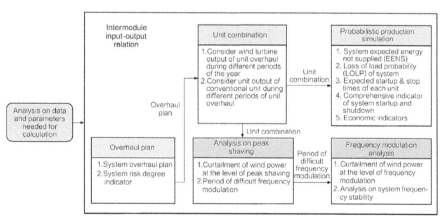

ATTACHED FIG. IV.1

The principle structure of the random production simulation model of the electric power system

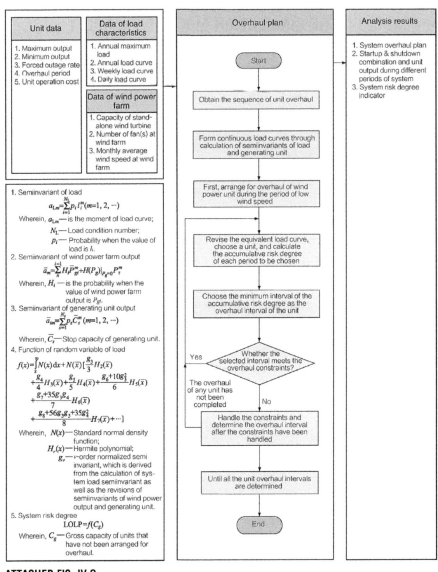

ATTACHED FIG. IV.2

The structure and process of the unit overhaul plan

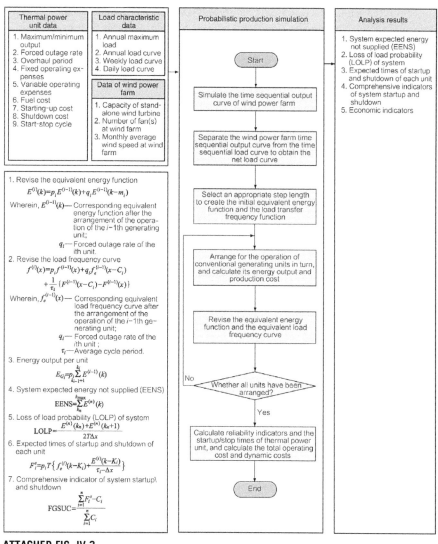

ATTACHED FIG. IV.3

The structure and process of random production simulation

APPENDIX V. THE MODEL FOR ANALYSIS ON FREQUENCY MODULATION OF THE ELECTRIC POWER SYSTEM

Analysis target: meet the requirements on continuous and steady-state system operation in a safe and economical manner.

Main constraints: minute-level electric power balance in key periods; system frequency; voltage of main nodes; maximum transmission capacity of tie lines; range and rate of power fluctuation of tie lines; regional reserves for load and accident; output rate of unit upgrade and downgrade; scope of unit output changes; other regional operations of various power stations; safety and stability (based on software including Bonneville Power Administration (BPA) Power System Analysis Software), etc.

Output result: minute-level continuous and steady-state operation analysis results in key periods, including regional system frequency modulation capacity surplus in key periods, system frequency, cross-regional power, change of node voltage, safe and sable operating margin of system, changes of unit output, etc.

The principle structure of the electric power system frequency modulation analysis model is shown in Attached Fig. V.1.

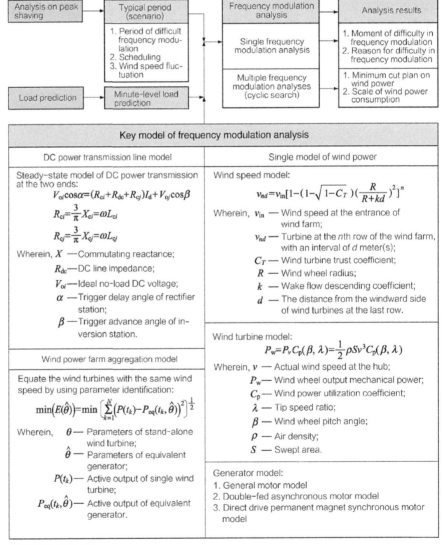

ATTACHED FIG. V.1

The principle structure of the frequency modulation analysis model of the electric power system

APPENDIX VI. COMPUTABLE GENERAL EQUILIBRIUM (CGE) MODEL

CGE model is an application of the general equilibrium theory in the real economy. It is based on the thought that various behavior subjects meet the assumptions of economic behavior, that is, the producers achieve the target of maximizing their profits by minimizing the costs under the definite condition of production technologies; the consumers maximize their consumption effects under the condition of a certain income level by making choices for consumption in accordance with their preferences; for import and export commodities as well as domestic products, the target of revenue maximization of domestic and foreign sales is realized through the price transmission mechanism under the condition of certain total output; and the supply of and demand for factors are the optimal arrangements for resource factor endowments in the process of production. The above behavioral assumptions are embodied in the production module, the revenue/expenditure module, and the commodity trade module of the CGE model.

The basic structure of the CGE model is shown in Attached Fig. VI.1. Generally, the model consists of three parts: (1) supply: the equation in this part mainly describes the behaviors of commodity and factor producers, and the conditions of

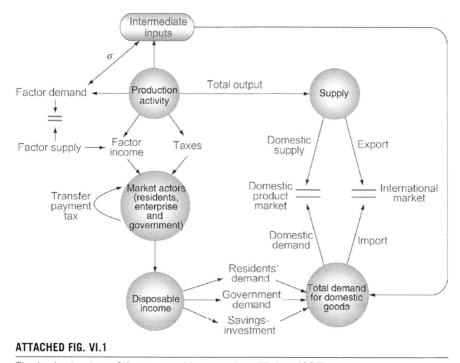

ATTACHED FIG. VI.1

The basic structure of the computable general equilibrium (CGE) model optimization, it covers the production equation, constraint equation, supply equation of production factors, and equation of optimization conditions, etc. (2) Demand: the total demand is generally divided into final consumption, intermediate input, and investment goods, and consumers are divided into household, enterprise, and government. The model mainly describes the behavior of consumers and conditions of optimization, including the consumer demand equation, constraint equation, production factor demand equation, intermediate demand equation, optimization condition equation, etc. In the open economic model, the CGE model needs to take the import and export models into account, and its consumption demand function allows replacement between import commodity and domestic goods. (3) Seek and establish the relationship between supply and demand. The basic idea is to seek a price vector so as to strike a supply-demand balance, and the market is the main channel linking the suppliers and those making demands. This part mainly describes market equilibrium and related budgetary balance. It includes product market equilibrium, factor market equilibrium, residents' income and expenditure balance, government budget balance, international market equilibrium, etc.

APPENDIX VII. THE INDICATOR SYSTEM FOR EVALUATION OF ENERGY DEVELOPMENT QUALITY

Energy development refers to, within a certain period, national and regional energy resources, production, conversion, transportation, consumption, import, and export, as well as balance and cross-regional dispatching of energy (by type and region). The core factors of energy development are "total quantity, structure, distribution, flow direction, and way of utilization," of which the flow direction involves the type, scale, and direction of energy flow. Energy development quality refers to the quality status that supports the energy changes of the economic and social development in different periods. Energy development benefits can be evaluated from the four aspects of "safety, economy, cleanliness, and efficiency."

Generally, the energy development quality evaluation indicator system consists of 48 subdivided indicators in 16 categories at the three levels of basic energy data, analytical indicator, and evaluation indicator. By building the development scenarios, the system defines basic energy data, measures analytical indicators, and conducts comprehensive evaluations on energy development benefits from the perspective of coordinated and sustainable development.

The basic energy data involve energy resources, energy production, energy conversion, energy transportation, energy consumption, and energy import and export, see Annexed Table VII.1. Specifically, the historical data come from various statistical yearbooks, and further development data come from the output results of the abovementioned integrated energy balance and optimization model.

Annexed Table VII.1 The basic dataset of energy development quality

Category	Name	Descriptions
Energy resources	Conditions of energy resources	Including resource reserves, structure, distribution and potential
	Development technology and cost	
	Energy price	
	Energy policy orientation	
Energy production	Production	Total quantity and type
	Distribution of production	Production of province (region)
Energy conversion	Way of conversion	By type
	Conversion efficiency	
Energy transportation	Way and scale of transportation	By type
	Transportation corridor and capacity	
Energy consumption	Consumption	Total quantity and type
	Distribution of consumption	Consumption of province (region)
Energy import and export	Total import and export	
	Import and export structure	
	International cooperation	

Based on statistics, analysis, and extraction of basic energy data, analytical indicators are used to describe the development features of the energy distribution. The indicators are categorized into energy efficiency, energy structure, energy supply and demand, energy dispatching, energy security, and energy environment, see Annexed Table VII.2 for details.

In analytical indicators, the principle of the center of gravity in physical science is used to create a concept of "energy moment," serving as a comprehensive indicator for measuring the energy transfer and large-scale distribution of resources. The physical meaning of physical energy moment is shown in Attached Fig. VII.1. Energy moment is defined as the product of cross-regional dispatching and the shipment distance, and its value reflects the dispatching demand arising from unbalanced regional energy production and consumption. Energy moment is the product of the outward energy transport volume from sending-end regions and corresponding shipment distance. Its computational formula is shown as follows:

Energy moment $= \displaystyle\sum_{i} \sum_{j}$ (the energy transport volume from the sending-end i

to the receiving-end j × the shipment distance from the sending-end i **to the receiving-end** j)

Annexed Table VII.2 Analytical indicators of energy development quality

Category	Name	Descriptions
Energy efficiency	Energy consumption per unit of GDP	Energy intensity
	Elastic coefficient of energy consumption	
	Per-capita energy consumption	
	Efficiency of energy utilization	
Energy structure	Consumption structure of primary energy	Including the proportion of non-fossil energy use to primary energy consumption
	Consumption structure of end-use energy	Including the proportion of power consumption to end-use energy consumption
	Power-generating energy structure	Including the proportion of coal for power generation to coal consumption
	Production structure of primary energy	
Energy supply and demand	Concentration ratio of energy production	=energy output of main energy-producing areas/total quantity of energy production (total quantity and type)
	Concentration ratio of energy consumption	=energy consumption of main energy-consuming areas/total energy consumption (total quantity and type)
	Difference between energy production and consumption in provinces (regions)	=energy output − energy consumption (total quantity and type)
	Ratio between energy production and consumption in provinces (regions)	=energy output/energy consumption
	Key areas of energy production/consumption	By using the standard method for calculating the focus of an irregular planar object, the key areas of energy production and consumption of China can be worked out, on the basis of regional average energy production and consumption per unit area in relation to quality density.

Annexed Table VII.2 Analytical indicators of energy development quality *–cont'd*

Category	Name	Descriptions
Energy dispatching	Total amount and structure of cross-regional energy dispatching	$=\sum$ Cross-regional energy dispatching from sending ends to receiving ends[a]
	Proportion of cross-regional energy dispatching	=cross-regional energy dispatching/total energy consumption (total quantity and type)
	Energy moment	$=\sum$ (Energy dispatching from sending ends to receiving ends × shipment distance)
	Average shipment distance of cross-regional energy dispatching	$=\sum$ (Energy dispatching from sending ends to receiving ends × shipment distance)/total cross-regional energy dispatching
	Utilization ratio of transportation corridor	=Shipping volume of transportation corridor/shipping capacity of transportation corridor
	Ratio of way of transportation	The ratio of energy resources transportation by different means, such as the ratio between coal transportation and power transmission
Energy security	External dependence	
	Capability of energy import	Total quantity and type
	Degree of satisfaction of total energy demand	
	Energy reserves	
Energy environment	Carbon dioxide emissions	
	Carbon emission intensity	
	Pollutant emission amount, emission intensity, and emission distribution	Total quantity and type, pollutants include sulfur dioxide, oxynitride, fume, and heavy metals, etc.
	Loss of energy environment	

[a]*It does not consider energy flows or regulation between different provinces (regions), but only considers the net energy import of China and the net amount of dispatching between different regions.*

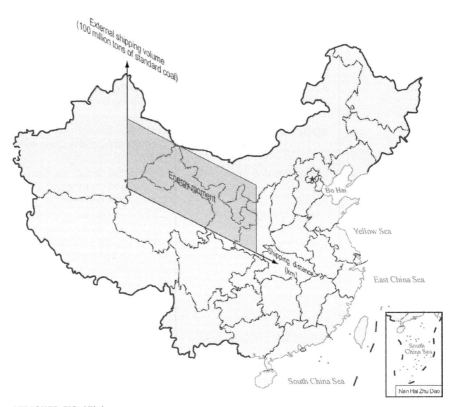

ATTACHED FIG. VII.1

The physical meaning of energy moment.

The average shipment distance is the average distance of energy transportation from major sending ends to major receiving ends, reflecting the distance of energy delivery. Its computational formula is shown further:

$$Average\ Shipment\ Distance = \frac{\sum_{Sending\ end\ i}\sum_{Receiving\ end\ j}\left(Volume_{i-j} \times Distance_{i-j}\right)}{\sum_{Sending\ end\ i}\sum_{Receiving\ end\ j}Volume_{i-j}}$$

where, $Volume_{i-j}$ is the energy transport volume from the sending end i to the receiving end j; $Distance_{i-j}$ is the shipment distance from the sending end i to the receiving end j

Evaluation indicators are established from the four aspects of safety, economy, cleanliness, and efficiency and designed for integrated and coordinated sustainable development, and achieving clear evaluation on the energy situation. The indicators for evaluation on energy development quality are shown in Annexed Table VII.3.

Annexed Table VII.3 Indicators for evaluation on energy development quality

Target	Category	Descriptions	Support indicators
Comprehensive coordination of sustainable development level	Safety	Evaluation of energy safety level	Energy structure, external dependence, import capability, energy reserves, total amount and proportion of energy dispatching, ratio of way of transportation, utilization ratio of transportation corridor, level of supply-demand balance, ecological security
	Economical efficiency	Evaluation of development cost effectiveness	Energy development technology and costs, energy price, economical efficiency of transportation, loss of energy environment
	Cleanliness	Evaluation of cleanliness development level	Proportion of non-fossil energy, pollutant emission, emission intensity, carbon dioxide emission, emission intensity
	Efficiency	Evaluation of comprehensive energy efficiency	Energy consumption per unit of GDP, elastic coefficient of energy consumption, efficiency of various processes (production, conversion, transportation, and utilization)

ANNEXED TABLES. DATA TABLES OF ENERGY PRODUCTION AND CONSUMPTION IN THE WORLD

Annexed Table A.1. Total consumption of primary energy in the world

Year	Primary energy consumption (one million tons of standard oil)	Year-on-year growth (%)
1965	3750	–
1966	3968	5.8
1967	4119	3.8
1968	4366	6.0
1969	4656	6.6
1970	4944	6.2
1971	5137	3.9
1972	5412	5.4
1973	5717	5.6

Continued

Annexed Table A.1. Total consumption of primary energy in the world –*cont'd*

Year	Primary energy consumption (one million tons of standard oil)	Year-on-year growth (%)
1974	5740	0.4
1975	5766	0.5
1976	6082	5.5
1977	6298	3.5
1978	6494	3.1
1979	6710	3.3
1980	6631	−1.2
1981	6583	−0.7
1982	6558	−0.4
1983	6647	1.4
1984	6969	4.8
1985	7161	2.8
1986	7326	2.3
1987	7575	3.4
1988	7853	3.7
1989	8021	2.1
1990	8105	1.0
1991	8146	0.5
1992	8189	0.5
1993	8246	0.7
1994	8352	1.3
1995	8564	2.5
1996	8792	2.7
1997	8903	1.3
1998	8968	0.7
1999	9127	1.8
2000	9356	2.5
2001	9434	0.8
2002	9614	1.9
2003	9950	3.5
2004	10,450	5.0
2005	10,754	2.9
2006	11,048	2.7
2007	11,348	2.7
2008	11,493	1.3
2009	11,391	−0.9
2010	11,978	5.1
2011	12,275	2.5
2012	12,634	1.4
2013	12,866	1.8
2014	12,989	1.0

From: BP, *Statistical Review of World Energy 2017*.

Annexed Table A.2. The consumption structure of primary energy in the world

Year	Total	Coal Total amount (1 million tons of standard oil)	Coal Proportion (%)	Petroleum Total amount (1 million tons of standard oil)	Petroleum Proportion (%)	Natural gas Total amount (1 million tons of standard oil)	Natural gas Proportion (%)	Non-fossil energy Total amount (1 million tons of standard oil)	Non-fossil energy Proportion (%)
1965	3750	1427	38.1	1513	40.3	594	15.8	216	5.8
1966	3968	1444	36.4	1645	41.5	646	16.3	233	5.9
1967	4119	1422	34.5	1764	42.8	692	16.8	241	5.8
1968	4366	1443	33.1	1915	43.9	753	17.2	254	5.8
1969	4656	1475	31.7	2082	44.7	827	17.8	271	5.8
1970	4944	1499	30.3	2260	45.7	898	18.2	287	5.8
1971	5137	1490	29.0	2383	46.4	958	18.7	306	6.0
1972	5412	1511	27.9	2567	47.4	1006	18.6	328	6.1
1973	5717	1553	27.2	2769	48.4	1052	18.4	343	6.0
1974	5740	1552	27.0	2727	47.5	1076	18.8	385	6.7
1975	5766	1587	27.5	2696	46.8	1071	18.6	412	7.1
1976	6082	1650	27.1	2870	47.2	1133	18.6	429	7.0
1977	6298	1704	27.1	2966	47.1	1167	18.5	461	7.3
1978	6494	1730	26.6	3046	46.9	1213	18.7	505	7.8
1979	6710	1799	26.8	3096	46.1	1288	19.2	527	7.9
1980	6631	1804	27.2	2979	44.9	1296	19.5	552	8.3
1981	6583	1818	27.6	2868	43.6	1309	19.9	588	8.9
1982	6558	1845	28.1	2777	42.3	1312	20.0	624	9.5
1983	6647	1894	28.5	2755	41.4	1328	20.0	670	10.1
1984	6969	1981	28.4	2815	40.4	1439	20.7	734	10.5
1985	7161	2062	28.8	2816	39.3	1488	20.8	796	11.1
1986	7326	2094	28.6	2901	39.6	1503	20.5	829	11.3
1987	7575	2172	28.7	2956	39.0	1579	20.8	869	11.5
1988	7853	2234	28.5	3047	38.8	1654	21.1	918	11.7

Continued

Annexed Table A.2. The consumption structure of primary energy in the world –*cont'd*

Year	Total	Coal		Petroleum		Natural gas		Non-fossil energy	
		Total amount (1 million tons of standard oil)	Proportion (%)	Total amount (1 million tons of standard oil)	Proportion (%)	Total amount (1 million tons of standard oil)	Proportion (%)	Total amount (1 million tons of standard oil)	Proportion (%)
1989	8021	2252	28.1	3103	38.7	1728	21.5	938	11.7
1990	8105	2207	27.2	3158	39.0	1769	21.8	971	12.0
1991	8146	2176	26.7	3157	38.8	1808	22.2	1005	12.3
1992	8189	2152	26.3	3208	39.2	1818	22.2	1011	12.3
1993	8246	2159	26.2	3175	38.5	1854	22.5	1056	12.8
1994	8352	2174	26.0	3240	38.8	1866	22.3	1073	12.8
1995	8564	2232	26.1	3280	38.3	1927	22.5	1125	13.1
1996	8792	2271	25.8	3350	38.1	2017	22.9	1154	13.1
1997	8903	2287	25.7	3436	38.6	2016	22.6	1163	13.1
1998	8968	2276	25.4	3460	38.6	2048	22.8	1184	13.2
1999	9127	2291	25.1	3526	38.6	2097	23.0	1214	13.3
2000	9356	2372	25.4	3572	38.2	2174	23.2	1233	13.2
2001	9434	2381	25.2	3596	38.1	2215	23.5	1242	13.2
2002	9614	2443	25.4	3630	37.8	2271	23.6	1270	13.2
2003	9950	2638	26.5	3703	37.2	2347	23.6	1263	12.7
2004	10,450	2839	27.2	3857	36.9	2418	23.1	1336	12.8
2005	10,754	2982	27.7	3902	36.3	2498	23.2	1373	12.8
2006	11,048	3139	28.4	3944	35.7	2549	23.1	1416	12.8
2007	11,348	3267	28.8	4005	35.3	2646	23.3	1430	12.6
2008	11,493	3324	28.9	3987	34.7	2712	23.6	1469	12.8
2009	11,391	3347	29.4	3909	34.3	2644	23.2	1492	13.1
2010	11,978	3532	29.5	4032	33.7	2843	23.7	1571	13.1
2011	12,275	3724	30.3	4059	33.1	2906	23.7	1586	12.9
2012	12,634	3817	30.2	4176	33.1	3011	23.8	1629	12.9
2013	12,866	3887	30.2	4221	32.8	3054	23.7	1703	13.2
2014	12,989	3889	29.9	4255	32.8	3073	23.7	1771	13.6

From: BP, Statistical Review of World Energy 2017.

Annexed Table A.3. The distribution of primary energy consumption in the world (unit: million tons of standard oil)

Year	Total	By region						By organization	
		North America	Central and South America	Europe	Middle East	Africa	Asia Pacific	OECD	Non-OECD
1965	3750	1428	109	1662	57	57	436	2625	1125
1966	3968	1510	117	1747	60	60	475	2769	1199
1967	4119	1563	122	1810	63	62	499	2879	1239
1968	4366	1660	128	1906	66	66	540	3072	1293
1969	4656	1751	135	2016	70	67	616	3283	1372
1970	4944	1817	142	2134	74	73	704	3465	1479
1971	5137	1856	150	2207	79	78	768	3554	1583
1972	5412	1953	162	2316	85	83	813	3725	1687
1973	5717	2036	178	2440	93	89	881	3932	1785
1974	5740	2000	187	2463	99	93	898	3871	1869
1975	5766	1954	189	2494	98	99	932	3790	1977
1976	6082	2060	202	2627	106	108	979	3995	2087
1977	6298	2117	217	2701	115	113	1035	4084	2214
1978	6494	2127	228	2797	124	118	1100	4146	2348
1979	6710	2162	242	2882	146	129	1151	4256	2454
1980	6631	2107	252	2825	138	144	1166	4146	2485
1981	6583	2057	250	2798	146	160	1172	4049	2533
1982	6558	1987	256	2801	159	172	1183	3939	2619
1983	6647	1973	259	2829	173	176	1237	3936	2711
1984	6969	2081	270	2917	195	183	1324	4123	2846
1985	7161	2085	275	3022	207	190	1383	4189	2973
1986	7326	2090	290	3084	221	195	1447	4240	3086
1987	7575	2162	300	3159	230	200	1525	4362	3214
1988	7853	2257	310	3200	239	214	1633	4508	3345
1989	8021	2319	316	3194	255	215	1723	4598	3423
1990	8105	2326	326	3195	266	220	1771	4630	3475
1991	8`46	2329	333	3137	276	222	1848	4673	3473

Continued

Annexed Table A.3. The distribution of primary energy consumption in the world (unit: million tons of standard oil) —cont'd

| Year | Total | By region | | | | | | By organization | |
		North America	Central and South America	Europe	Middle East	Africa	Asia Pacific	OECD	Non-OECD
1992	8189	2369	347	3036	294	222	1921	4723	3466
1993	8246	2418	360	2919	306	225	2019	4795	3451
1994	8352	2472	380	2802	332	234	2132	4883	3469
1995	8564	2517	397	2783	347	244	2276	4999	3565
1996	8792	2604	416	2801	360	253	2358	5170	3622
1997	8903	2628	436	2763	377	259	2440	5224	3679
1998	8968	2643	452	2774	394	264	2440	5244	3724
1999	9127	2690	455	2771	407	271	2533	5318	3810
2000	9356	2757	466	2810	421	274	2627	5435	3920
2001	9434	2699	468	2852	445	280	2690	5407	4027
2002	9614	2739	473	2849	465	288	2800	5448	4165
2003	9950	2756	479	2913	486	302	3014	5507	4443
2004	10,450	2818	502	2958	523	321	3328	5622	4828
2005	10,754	2839	522	2969	562	327	3535	5669	5086
2006	11,048	2819	546	3009	587	331	3755	5674	5375
2007	11,348	2869	568	3003	612	350	3947	5718	5629
2008	11,493	2819	587	3007	652	368	4060	5661	5832
2009	11,391	2687	583	2831	671	366	4254	5389	6003
2010	11,978	2764	619	2939	716	382	4558	5572	6405
2011	12,275	2773	643	2923	748	385	4803	5528	6747
2012	12,634	2724	681	2936	781	403	5109	5482	7152
2013	12,866	2796	697	2901	812	415	5245	5540	7326
2014	12,989	2821	704	2838	840	428	5357	5493	7491

From: BP, Statistical Review of World Energy 2017.

Annexed Table A.4. The coal consumption in the world (raw coal) (unit: million tons)

Year	Total	North America	Central and South America	Europe	Middle East	Africa	Asia Pacific	OECD	Non-OECD
				By region				By organization	
1965	2854	616	12	1704	0	56	466	1761	1093
1966	2887	644	12	1687	0	55	489	1750	1137
1967	2843	632	13	1664	1	57	477	1719	1125
1968	2886	656	12	1669	1	60	487	1761	1124
1969	2951	657	13	1671	1	60	548	1783	1168
1970	2999	654	14	1647	1	62	621	1765	1234
1971	2980	619	13	1618	1	67	662	1670	1309
1972	3022	642	13	1614	1	67	684	1651	1371
1973	3107	688	13	1630	1	71	703	1722	1385
1974	3104	673	14	1618	2	74	722	1703	1400
1975	3173	673	14	1629	1	79	778	1681	1492
1976	3300	724	15	1680	1	83	796	1769	1531
1977	3409	753	17	1700	1	85	854	1805	1604
1978	3460	737	18	1725	1	82	897	1789	1672
1979	3597	799	19	1750	1	87	942	1890	1707
1980	3608	826	20	1667	1	93	1000	1948	1660
1981	3636	851	20	1627	1	110	1027	1966	1669
1982	3689	825	21	1655	2	121	1066	1952	1737
1983	3789	856	23	1653	3	122	1132	1985	1804
1984	3952	921	27	1646	5	131	1232	2077	1885
1985	4125	940	31	1702	5	135	1312	2163	1962
1986	4187	926	31	1717	6	138	1369	2144	2043
1987	4343	968	32	1737	6	142	1459	2196	2148
1988	4469	1014	33	1703	6	153	1560	2246	2223
1989	4504	1025	34	1667	6	144	1627	2251	2253
1990	4414	1027	34	1583	7	149	1614	2201	2213

Continued

Annexed Table A.4. The coal consumption in the world (raw coal) (unit: million tons) –cont'd

Year	Total	North America	Central and South America	Europe	Middle East	Africa	Asia Pacific	OECD	Non-OECD
				By region				By organization	
1991	4352	1016	35	1476	7	145	1673	2159	2193
1992	4303	1024	34	1388	9	137	1712	2102	2202
1993	4317	1055	35	1283	10	143	1792	2092	2225
1994	4348	1063	36	1207	10	150	1883	2097	2252
1995	4464	1073	36	1163	11	157	2025	2112	2353
1996	4543	1122	38	1135	12	165	2070	2180	2363
1997	4575	1148	40	1096	13	171	2107	2194	2381
1998	4552	1164	39	1059	14	170	2106	2193	2359
1999	4581	1162	40	1008	13	166	2191	2167	2414
2000	4744	1213	40	1050	15	165	2260	2266	2479
2001	4762	1186	38	1037	17	164	2320	2248	2514
2002	4886	1183	37	1039	17	169	2441	2263	2624
2003	5275	1197	40	1073	18	179	2769	2312	2964
2004	5679	1203	42	1061	18	188	3167	2339	3340
2005	5965	1234	43	1027	18	183	3459	2362	3603
2006	6278	1208	42	1061	18	184	3765	2359	3919
2007	6535	1224	46	1066	19	193	3986	2404	4131
2008	6648	1202	49	1040	18	205	4135	2356	4292
2009	6693	1060	46	942	17	192	4436	2113	4580
2010	7064	1119	56	967	17	196	4709	2222	4842
2011	7449	1067	60	998	17	200	5106	2197	5251
2012	3011	820	144	967	374	109	599	1429	1581
2013	3054	844	149	949	396	111	606	1457	1597
2014	3073	862	152	905	415	114	625	1435	1638

Annexed Table A.5. The coal production in the world (raw coal) (unit: million tons)

Year	Total	By region						By organization	
		North America	Central and South America	Europe	Middle East	Africa	Asia Pacific	OECD	Non-OECD
1981	3836	790	11	1917	1	136	981	2045	1791
1982	3980	807	12	1973	1	149	1038	2103	1877
1983	3986	759	13	1964	1	151	1098	2044	1943
1984	4191	875	16	1938	1	168	1192	2143	2048
1985	4441	868	19	2060	1	179	1314	2262	2179
1986	4549	870	20	2111	1	183	1362	2298	2251
1987	4650	901	24	2119	1	184	1421	2326	2323
1988	4755	938	27	2117	1	189	1483	2342	2413
1989	4838	966	31	2065	1	184	1590	2356	2482
1990	4740	1009	30	1890	1	183	1628	2286	2454
1991	4557	981	31	1695	1	186	1662	2169	2388
1992	4519	977	33	1611	1	182	1715	2105	2414
1993	4396	933	32	1481	1	190	1759	1998	2398
1994	4484	1019	34	1365	1	204	1862	2044	2440
1995	4605	1021	37	1313	1	214	2019	2041	2564
1996	4680	1051	40	1278	1	214	2096	2072	2608
1997	4730	1078	45	1255	1	227	2124	2108	2622
1998	4652	1100	47	1191	1	233	2080	2092	2560
1999	4638	1081	46	1150	1	229	2131	2059	2579
2000	4701	1054	54	1173	1	231	2188	2028	2673
2001	4918	1105	58	1198	1	230	2325	2105	2813
2002	4961	1070	53	1171	1	226	2439	2070	2891

Continued

Annexed Table A.5. The coal production in the world (raw coal) (unit: million tons) *–cont'd*

| Year | Total | North America | By region | | | | | By organization | |
			Central and South America	Europe	Middle East	Africa	Asia Pacific	OECD	Non-OECD
2003	5314	1044	62	1200	1	243	2764	2047	3267
2004	5723	1084	67	1195	1	249	3126	2088	3635
2005	6049	1106	73	1201	1	250	3419	2117	3932
2006	6357	1133	80	1220	2	249	3673	2141	4215
2007	6588	1122	84	1234	2	252	3896	2148	4441
2008	6822	1143	88	1251	2	256	4082	2162	4660
2009	6905	1050	88	1180	1	254	4332	2058	4847
2010	7255	1063	89	1194	1	259	4648	2060	5194
2011	7695	1077	101	1257	1	260	5000	2082	5613
2012	8208	1005	99	1305	1	267	5530	2058	6150
2013	8275	976	98	1257	2	268	5673	2027	6248
2014	8198	989	102	1207	1	276	5621	2057	6141

From: BP, Statistical Review of World Energy 2017.

Annexed Table A.6. The petroleum consumption in the world (unit: million tons)

Year	Total	By region						By organization	
		North America	Central and South America	Europe	Middle East	Africa	Asia Pacific	OECD	Non-OECD
1965	1513	620	81	574	48	26	163	1128	385
1966	1645	651	87	642	49	29	187	1228	418
1967	1764	678	90	697	51	29	219	1317	447
1968	1915	725	95	760	53	30	251	1436	480
1969	2082	765	100	838	55	31	292	1567	515
1970	2260	798	104	928	58	35	338	1697	563
1971	2383	825	110	975	62	37	375	1778	605
1972	2567	891	119	1047	67	40	403	1908	659
1973	2769	941	130	1123	72	43	460	2049	719
1974	2727	912	134	1105	75	45	456	1963	764
1975	2696	898	134	1093	73	47	453	1904	792
1976	2870	963	140	1154	80	52	481	2031	839
1977	2966	1012	148	1159	88	55	504	2079	887
1978	3046	1000	154	1209	94	58	531	2098	948
1979	3096	989	160	1235	107	61	545	2110	986
1980	2979	929	163	1198	103	69	517	1965	1014
1981	2868	874	160	1149	111	73	502	1847	1021
1982	2777	826	158	1110	120	77	486	1748	1029
1983	2755	815	153	1085	130	80	491	1718	1037
1984	28˙5	843	153	1089	139	81	509	1764	1051
1985	28˙6	843	150	1086	148	84	505	1744	1071
1986	2901	873	157	1113	150	82	525	1803	1098
1987	2956	894	161	1118	156	85	542	1828	1128
1988	3047	929	165	1121	155	89	588	1896	1151
1989	31C3	936	166	1120	163	93	625	1920	1183

Continued

Annexed Table A.6. The petroleum consumption in the world (unit: million tons) –cont'd

Year	Total	By region						By organization	
		North America	Central and South America	Europe	Middle East	Africa	Asia Pacific	OECD	Non-OECD
1990	3158	923	170	1130	175	95	664	1940	1218
1991	3157	905	171	1108	182	98	693	1945	1212
1992	3208	924	182	1077	187	99	738	1995	1213
1993	3175	930	184	999	191	100	770	2003	1172
1994	3240	959	195	962	205	103	816	2060	1179
1995	3280	951	201	949	210	107	862	2076	1204
1996	3350	984	206	936	216	108	899	2135	1214
1997	3436	1001	217	942	220	112	944	2167	1269
1998	3460	1023	226	947	228	114	922	2174	1286
1999	3526	1047	229	938	235	118	959	2208	1318
2000	3572	1060	227	929	243	118	994	2217	1355
2001	3596	1061	231	938	249	120	996	2218	1377
2002	3630	1060	229	936	257	122	1025	2210	1420
2003	3703	1082	223	945	268	125	1061	2243	1460
2004	3857	1126	232	955	287	131	1126	2286	1571
2005	3902	1131	238	959	297	137	1140	2302	1600
2006	3944	1121	244	973	309	136	1161	2291	1654
2007	4005	1125	259	952	323	143	1203	2277	1728
2008	3987	1070	269	956	342	150	1202	2209	1778
2009	3909	1019	266	908	350	154	1211	2098	1811
2010	4032	1041	281	903	364	161	1282	2118	1914
2011	4059	1026	289	898	371	158	1316	2092	1967
2012	4176	1012	321	882	389	169	1403	2072	2104
2013	4221	1025	332	864	399	175	1425	2059	2162
2014	4255	1027	337	859	408	178	1447	2037	2218

From: BP, Statistical Review of World Energy 2017.

Annexed Table A.7. The petroleum production in the world (unit: million tons)

| Year | Total | By region | | | | | | By organization | | | |
		North America	Central and South America	Europe	Middle East	Africa	Asia Pacific	OECD	Non-OECD	OPEC	Non-OPEC
1965	1568	490	226	282	419	107	45	516	1052	699	626
1966	1702	521	224	305	467	135	50	547	1155	770	668
1967	1827	557	238	329	500	149	53	584	1242	819	719
1968	1993	582	245	351	562	191	62	610	1383	912	772
1969	2143	597	245	370	617	243	71	625	1518	1007	808
1970	2358	628	252	395	692	292	99	662	1696	1132	873
1971	2496	625	246	419	814	274	119	666	1831	1229	890
1972	2641	640	234	444	908	275	140	682	1959	1317	924
1973	2871	641	249	472	1055	287	167	687	2184	1489	953
1974	2879	618	227	503	1089	265	178	664	2215	1476	945
1975	2738	592	192	543	980	242	189	647	2091	1299	948
1976	2973	578	190	587	1113	289	216	649	2324	1471	982
1977	3077	593	189	639	1118	303	235	690	2387	1485	1047
1978	3137	626	189	683	1065	298	246	742	2364	1418	1117
1979	3237	644	202	723	1089	326	252	787	2451	1482	1170
1980	3092	671	195	747	935	301	245	820	2272	1287	1201
1981	2914	682	193	760	797	239	242	840	2074	1079	1226
1982	2800	706	185	781	660	231	237	881	1919	920	1268
1983	2763	709	183	805	578	233	254	904	1859	829	1318
1984	2819	728	193	817	551	249	280	946	1873	808	1398
1985	2797	730	193	807	516	261	289	959	1838	772	1428
1986	2942	705	206	830	639	261	300	935	2006	902	1425
1987	2953	701	204	844	641	260	303	937	2015	890	1437
1988	3075	697	214	841	742	275	306	931	2144	1008	1443

Continued

Annexed Table A.7. The petroleum production in the world (unit: million tons) –cont'd

Year	Total	By region						By organization			
		North America	Central and South America	Europe	Middle East	Africa	Asia Pacific	OECD	Non-OECD	OPEC	Non-OPEC
1989	3109	665	216	819	797	297	315	892	2217	1077	1424
1990	3175	655	234	788	852	321	326	894	2282	1159	1446
1991	3166	669	247	745	841	328	335	919	2246	1164	1485
1992	3195	663	252	696	914	335	335	929	2267	1242	1502
1993	3194	653	262	660	952	332	337	928	2266	1274	1518
1994	3244	648	278	663	975	334	346	969	2275	1301	1579
1995	3286	646	300	669	979	339	352	976	2310	1317	1610
1996	3384	660	321	680	1003	356	364	1008	2376	1365	1666
1997	3486	670	337	689	1051	370	368	1021	2465	1433	1691
1998	3551	667	356	686	1112	364	366	1013	2538	1492	1697
1999	3487	639	345	700	1081	360	363	991	2496	1435	1681
2000	3618	651	352	725	1141	371	380	1014	2604	1511	1714
2001	3607	652	346	747	1112	374	376	1003	2604	1477	1705
2002	3588	660	341	786	1044	380	377	1009	2579	1402	1719
2003	3705	670	325	819	1118	399	373	999	2705	1485	1706
2004	3879	668	340	850	1197	447	378	979	2900	1621	1699
2005	3916	646	358	845	1216	473	380	933	2983	1680	1659
2006	3929	647	359	848	1225	472	377	913	3016	1689	1639
2007	3929	641	359	860	1204	486	379	898	3031	1679	1625
2008	3965	618	366	851	1258	488	384	864	3101	1737	1601
2009	3869	632	372	857	1167	463	379	864	3005	1614	1611
2010	3945	651	375	854	1191	479	396	868	3077	1646	1641
2011	3996	670	380	839	1301	417	388	867	3129	1696	1640
2012	4116	720	379	834	1344	440	400	902	3214	1780	2336
2013	4125	784	379	833	1326	409	394	954	3171	1732	2393
2014	4226	869	393	835	1339	394	397	1042	3184	1730	2496

From: BP, Statistical Review of World Energy 2017.

Annexed Table A.8. The natural gas consumption in the world (unit: billion m³)

Year	Total	By region						By organization	
		North America	Central and South America	Europe	Middle East	Africa	Asia Pacific	OECD	Non-OECD
1965	651	464	14	156	10	1	6	490	161
1966	708	500	15	175	11	1	6	530	178
1967	759	526	16	196	12	1	7	563	196
1968	825	565	17	222	13	1	8	615	211
1969	903	611	18	252	15	1	11	678	229
1970	985	645	18	290	16	2	15	736	250
1971	1053	667	19	331	17	2	17	781	272
1972	1108	681	20	367	18	2	20	822	287
1973	1162	683	22	408	21	4	25	847	315
1974	1189	661	23	445	24	4	31	850	338
1975	1185	615	23	480	25	5	36	815	369
1976	1254	627	27	526	26	6	41	847	407
1977	1291	618	29	558	28	7	51	850	441
1978	1343	625	30	589	30	10	58	872	472
1979	1426	647	33	624	40	15	66	913	513
1980	1436	638	35	636	35	20	72	908	528
1981	1450	625	36	656	36	24	74	892	558
1982	1453	591	40	681	39	26	77	852	601
1983	1470	556	42	717	43	28	85	826	644
1984	1592	593	45	771	56	27	100	887	706
1985	1646	576	46	825	61	29	110	884	763
1986	1665	543	50	847	73	33	120	858	807
1987	1748	572	50	889	76	34	127	905	842
1988	1832	601	54	916	87	37	136	934	897
1989	1913	636	56	941	96	39	145	985	928
1990	1959	638	58	974	96	40	155	1000	959

Continued

Annexed Table A.8. The natural gas consumption in the world (unit: billion m³) –cont'd

Year	Total	North America	By region						By organization	
			Central and South America	Europe	Middle East	Africa	Asia Pacific	OECD	Non-OECD
1991	2002	651	58	986	98	40	168	1034	967
1992	2013	674	60	949	111	43	177	1059	954
1993	2054	694	65	946	119	43	187	1099	955
1994	2066	712	68	906	132	45	204	1126	940
1995	2135	743	75	915	142	47	213	1187	948
1996	2235	760	83	957	151	50	234	1251	984
1997	2234	767	85	921	165	49	247	1263	971
1998	2267	753	90	944	175	51	254	1267	1000
1999	2323	760	90	967	181	55	270	1302	1021
2000	2409	795	96	983	187	58	291	1356	1053
2001	2454	760	101	1014	207	64	308	1341	1113
2002	2516	788	102	1017	218	66	325	1370	1145
2003	2599	779	108	1060	229	73	351	1394	1206
2004	2679	785	118	1083	247	80	367	1418	1261
2005	2767	777	123	1106	279	83	398	1426	1341
2006	2824	772	135	1112	291	88	425	1426	1399
2007	2930	814	135	1126	303	94	458	1477	1453
2008	3005	821	141	1131	332	100	480	1499	1506
2009	2931	810	135	1045	344	99	497	1451	1479
2010	3153	836	150	1125	377	107	558	1536	1617
2011	3223	864	155	1101	403	110	591	1535	1688
2012	3338	903	160	1074	415	121	665	1581	1757
2013	3384	928	165	1054	440	123	673	1609	1774
2014	3401	944	169	1006	461	127	694	1581	1820

Annexed Table A.9. The natural gas production in the world (unit: billion m^3)

Year	Total	By region						By organization	
		North America	Central and South America	Europe	Middle East	Africa	Asia Pacific	OECD	Non-OECD
1970	1001	663	18	282	20	3	16	746	256
1971	1066	685	19	318	22	5	19	790	277
1972	1118	694	20	351	25	7	21	824	294
1973	1171	703	22	383	28	9	26	851	320
1974	1192	673	23	423	32	10	31	839	353
1975	1195	633	24	457	33	12	36	805	390
1976	1243	630	26	498	34	14	41	813	430
1977	1293	636	28	527	36	15	50	826	467
1978	1339	636	30	557	39	19	58	832	507
1979	1427	659	31	595	47	30	65	864	563
1980	1434	650	34	618	38	24	70	852	582
1981	1457	643	35	642	39	26	71	841	616
1982	1459	610	38	663	43	31	74	797	662
1983	1465	557	42	698	46	42	80	749	716
1984	1596	602	45	749	59	43	98	799	796
1985	1649	579	46	806	64	47	107	785	864
1986	1697	559	50	841	76	52	119	763	934
1987	1780	582	50	883	82	57	127	793	987
1988	1863	610	54	910	93	60	136	810	1054
1989	1923	621	56	935	103	66	142	829	1094
1990	1980	640	58	961	101	69	151	852	1129
1991	2002	643	60	956	105	74	164	868	1133
1992	2018	658	61	932	114	79	174	888	1130
1993	2057	680	65	923	123	82	185	923	1135

Continued

Annexed Table A.9. The natural gas production in the world (unit: billion m^3) –*cont'd*

Year	Total	North America	By region					By organization	
			Central and South America	Europe	Middle East	Africa	Asia Pacific	OECD	Non-OECD
1994	2083	713	69	886	136	78	201	964	1120
1995	2115	716	76	877	150	85	212	977	1138
1996	2215	732	84	922	158	91	227	1027	1188
1997	2220	738	85	877	176	102	242	1031	1188
1998	2269	750	90	891	185	108	246	1043	1226
1999	2331	748	92	913	195	120	263	1050	1281
2000	2411	764	100	937	208	130	272	1074	1337
2001	2477	780	105	945	233	131	282	1097	1380
2002	2519	763	107	966	247	134	301	1087	1432
2003	2617	767	119	1001	263	145	322	1094	1523
2004	2688	755	132	1025	285	155	336	1094	1595
2005	2770	746	139	1029	320	174	363	1079	1692
2006	2869	764	151	1042	339	191	382	1092	1778
2007	2939	782	152	1043	358	203	400	1101	1838
2008	3047	801	158	1075	384	212	417	1131	1916
2009	2956	803	152	955	407	199	440	1122	1834
2010	3178	819	163	1027	472	214	484	1148	2030
2011	3276	864	168	1036	526	203	479	1168	2108
2012	3352	879	173	1026	555	214	505	1197	2155
2013	3404	885	176	1033	587	206	517	1202	2202
2014	3466	937	177	1003	603	207	539	1248	2218

From: BP, Statistical Review of World Energy 2017.

Bibliography

[1] UNFPA, State of world population report 2015, 2015.
[2] IMF, World economic outlook 2015, 2015.
[3] BP, Statistical review of world energy 2015, 2015.
[4] IEA, Key world statistics 2015, 2015.
[5] EIA, Annual energy outlook 2015, 2015.
[6] IEA, Technology roadmap solar photovoltaic energy, 2014.
[7] BP, Energy outlook, 2016.
[8] K. Porter, J. Rogers, R. Wiser, Update on wind curtailment in Europe and North America, Consultants to the Center for Resource Solutions, June 16, 2011.
[9] China Statistical Yearbook 2016, National Bureau of Statistics of China, 2016.
[10] Statistics Data of National Electric Power Industry 2012, China Electricity Council, 2012.
[11] Statistics Data of National Electric Power Industry 2013, China Electricity Council, 2013.
[12] Statistics Data of National Electric Power Industry 2014, China Electricity Council, 2014.
[13] Statistics Data of National Electric Power Industry 2015, China Electricity Council, 2015.
[14] Statistics Data of National Electric Power Industry 2016, China Electricity Council, 2016.
[15] IEA, Key world statistics 2015, 2015.
[16] IEA, Energy balances of OECD countries 2015, 2015.
[17] IEA, Energy balances of non-OECD countries 2015, 2015.
[18] IEA, Energy balance flows 2015, 2015.
[19] Analysis report on the supply and demand of power-generating energy and the development of power supply in China 2016, 2016.
[20] China Energy Statistics Yearbooks 2014, National Bureau of Statistics of China, 2014.
[21] China Energy Statistics Yearbooks 2015, National Bureau of Statistics of China, 2015.
[22] China Energy Statistics Yearbooks 2016, National Bureau of Statistics of China, 2016.
[23] Boxiong SHEN, Chen ZUO, Progress of research on mercury emission from coal combustion flue gas and control thereof, J. Saf. Environ. 2012 02.
[24] Review results of hydropower resources of the People's Republic of China 2005, 2005.
[25] China mineral processing technology network, http://www.miningl20.com.
[26] Nuclear power in China, http://www.world-nuclear.org.
[27] R.D. Wilson, R. Krakowski, Planning for future energy resources, Science (300) 2003 25 (Letter).
[28] Scientific development in 2030: research report on China's National Energy Strategy, 2010.
[29] Wind & Solar Energy Evaluation Center, China Meteorological Administration.
[30] China energy development report 2010 of the National Energy Administration, 2010.
[31] General research report on grid connection of market consumption of wind power of the National Energy Administration, 2011.
[32] National power industry statistical express 2015, 2016.
[33] Development of wind power industry in 2015, 2016.
[34] Research on the prospect of development and utilization of China's solar energy resources, 2009.
[35] 12th Five-Year Plan on Biomass Energy Development, National Energy Administration, 2012.

[36] A survey and evaluation of the potential of China's shale gas resources and favorable area optimization, 2012.

[37] The grid development plan of the state grid corporation of China for the 12th five-year plan period.

[38] The research report of southern China power grid on the electric power industrial development during the 12th five-year plan and on a medium and long term.

[39] Medium and Long-term Renewable Energy Planning, National Energy Administration of China, 2007.

[40] The "12th Five-Year Plan on Renewable Energy Development", National Energy Administration of China, 2012.

[41] Scientific development 2030 – research report on China's National Energy Strategy of National Energy Administration.

[42] Strategic Action Plan on Energy Development (2014–2020), National Energy Administration of China, 2014.

[43] 13th "Five-year Plan" on Energy Development, National Energy Administration of China, 2016.

[44] 13th "Five-year Plan" on Electric Power Development, National Energy Administration of China, 2016.

[45] Guidance catalogue for adjustment of industrial structure (2014) released by the National Development and Reform Commission, 2014.

[46] Three Gorges Hydropower Station, China Water & Power Press, Beijing, 2009.

[47] 12th "Five-Year Plan" on Energy Development, National Energy Administration of China, 2012.

Index

Note: Page numbers followed by *f* indicate figures, and *t* indicate tables.

A

Alternating current/direct current (AC/DC)
    power transmission, 204–205, 256,
        260–261
American Recovery and Reinvestment Act, 30–31

B

Badi hydropower station, 232–233, 235*f*, 236*t*
Baihetan hydropower station, 231, 231*f*, 232*t*
"Belt and Road" initiative, 132–133
Biological liquid fuel, 110
Biomass energy
    biological liquid fuel, 110
    biomass briquette fuels, 112
    biomass gas, 112
    biomass power generation, 110
    development planning, 113–114
    development potential, 112–113
    distribution, 110
    resources, 109, 111*t*
British Petroleum (BP), 38, 39*t*

C

Carbon dioxide emissions, 53–54, 77
Cascade hydropower stations
    in Dadu River Basin, 232–235
    in Jinsha River, 230–231
    in Yalong River Basin, 232
CFPPs. *See* Coal-fired power plants (CFPPs)
CGE model. *See* Computable general
        equilibrium (CGE) model
Changheba hydropower station, 233–234, 235*f*,
        236*t*
Chemical energy storage technologies, 228–229
Clean energy
    definition, 269
    power generation, 272
2030 Climate and Energy Policy Framework, 14, 16
Coal, 9, 9*f*
    clean development and utilization, 117–118, 254
    coal-fired power base, development potential of,
        116–117
    consumption
        thermal power generation, 271

total energy consumption, control measures,
        242
    world consumption, 295–296*t*
development of, 56
ecological environment, pollution of, 50–53
equivalent calculation method, 270–271
power-generating energy source, 259
power unit, techno-economic parameters of, 272,
        273*t*
prices of, 273, 274*t*
production, 297–298*t*
proportion of
    EUECS, 42–43, 43*f*
    PECS, 37–39, 38*f*
supply capability, 114–115
transportation, 48–49, 48*f*, 60
Coal-fired power plants (CFPPs), 49, 56, 216,
        220–221
Computable general equilibrium (CGE) model,
        215, 223, 283
    analysis process of, 73, 74*f*
    basic structure of, 283–284, 283*f*
    energy transportation and circulation, costs of, 76
    feature of, 73
    multiregional dynamic CGE model, 73–75
    pollutants formation, mechanism of, 77
    power investments, 75
    power sector, capital formation of, 75
Concentrating solar power (CSP) development,
        27–29, 28*f*
Control of total energy consumption (CTEC)
    electrification level, upgradation of, 55
    goals and methods of, 55
    proposal of, 54

D

Dadu River Basin, 232–235
Dagangshan hydropower station, 233–234, 235*f*,
        236*t*
Danba hydropower station, 232–233, 235*f*, 236*t*
Demand slowdown scenario, 138–139, 148
Development goals, non-fossil energy, 226*f*
    energy storage technologies, 228–229
    hydropower
        cascade hydropower stations, 230–235

Development goals, non-fossil energy *(Continued)*
challenges, 227
clean energy power generation, 227, 229
hydropower projects, approval of, 235–237
installed capacity, 227, 229–230, 234–235
large-scale hydropower construction,
capability of, 230
nuclear power
clean energy power generation, 227, 229
installed capacity, 227, 229
large-scale nuclear power development, 227,
237–239
nuclear power station construction,
demands for, 237
power grid development, 228
solar power
grid connection and consumption, 239–240
installed capacity of, 229
photothermal power generation projects,
demonstration of, 228
PV power generation technology, 239
rooftop grid connection systems, 228
total energy demand, control of, 225–226
wind power
grid connection and consumption, 239–240
installed capacity of, 229, 239
technical feasibility, problems of, 227–228
technological level and reliability of, 239
Direct current (DC) system, 203
Dispatching automation technology, 228

E
Electric car, 221
Electric power
development trend
global power-generating mix, 33, 33*f*
global wind power and solar power, 33–34
low-carbon green development, 32
major power failure, large-scale
synchronous power grids, 34–35, 35*f*
nonhydrorenewable energy, 34–35
power grid development, 34–35
power transmission network expansion, 36
smart micro-grid technology, 36
frequency modulation analysis model, 281–282,
282*f*
GESP-IV model, 275–278
random production simulation model, 279–280
Electrothermal equivalent calculation method,
270–271
End-user energy consumption structure (EUECS),
6, 7*t*, 42–43
Energy 2020, 14, 16

Energy conservation, 241–242, 253–254
Energy development
CGE model, 73–77
comprehensive energy transportation system,
59–61
control of total energy consumption, 54–55
energy structure adjustment and layout
optimization
coal and CFPPs, development of, 56
hydropower, 57–58
natural gas and gas-fired power generation,
56–57
nuclear power, 58
solar power, 59
wind power, 58–59
IO model, 66, 77–79
problems in
carbon emission reduction, pressure of, 53–54
conventional fossil energy, unsustainable
supply of, 46–47
ecological environment, destruction of, 50–53
energy transportation, 48–49, 48*f*
import energy, increased dependence on,
47–48
inefficient energy consumption, 44–46
peak-regulating and power transmission
ability, lack of, 49–50
quality, 284
analytical indicators, 285, 286–287*t*, 288, 288*f*
basic energy data, 284, 285*t*
evaluation indicators, 79–80, 79*f*, 288, 289*t*
research model, framework of, 63–65, 64*f*
Energy Independence and Security Act, 11
Energy Information Agency (EIA), 129–130
Energy loss, 270
Energy moment, 285, 288*f*
Energy resources
biomass energy
biological liquid fuel, 110
biomass briquette fuels, 112
biomass gas, 112
biomass power generation, 110
development planning, 113–114
development potential, 112–113
distribution, 110
resources, 109, 111*t*
coal *(see* Coal)
hydroenergy
development planning, 89–90
development potential, 89
installed hydropower capacity, 85–88, 87–88*f*
resources and distribution, 81–85
natural gas *(see* Natural gas)

nuclear energy
  development planning, 96–97
  development potential, 96
  development situation, 92–95
  uranium resource reserves and distribution,
    90–92
  oil, 118–122
  solar energy (*see* Solar energy resources)
  wind energy (*see* Wind energy resources)
Energy Roadmap 2050, 14, 16
Energy science and technological innovation,
    253–257
  coal development and utilization, 254
  energy efficiency, improvements in, 254
  energy-saving technology, improvements
    in, 254
  grid-connected power transmission and
    distribution, 257
  hydropower, 256
  natural gas power generation technologies, 255
  new energy power generation, 257
  nuclear power, 256
  power grid, 256
Energy storage technology, 228–229
  development roadmap, 30–31, 31*f*
  electrochemistry energy storage, 29–30
  physical energy storage, 29–30
  readiness level, 29–30, 30*f*
  superconductivity electromagnetism storage,
    29–30
Energy structure and non-fossil energy utilization
  end-user energy consumption structure, 42–43
  PECS
    evolution of, 37–38, 38*f*
    of major energy-consuming countries, 38–39,
      39*t*
  PEPS, 39
  power generation structure, 40–41
Energy transportation system, 59–61
Ertan hydropower station, 232, 233*f*, 234*t*
EUECS. *See* End-user energy consumption
    structure (EUECS)
EU Energy Strategy 2020–2050, 14
  2030 Climate and Energy Policy Framework,
    14–16
  Energy 2020, 14–15
  energy development goal, 14–15, 15*t*
  Energy Roadmap 2050, 14, 16
European–American power grid integration
  European power grid integration, 25–27, 26*f*
  North American Interconnected Power Grid,
    23–24, 24–25*f*
EU's Emission Trading Scheme (EUETS), 14–15

EU-Ten Year Network Development Plan
    (EU-TYNDP), 21
Expected development scenario
  boundary conditions, 141
  installed clean energy capacity
    biomass power, 143
    hydropower, 141–142
    nuclear power, 142
    solar power, 143
    wind power, 142–143
Extra high-voltage (EHV) power grids, 260

F
Final energy consumption, 270
Fixed-production pumped storage projects,
    272–273, 274*t*
Flow battery, 228–229
Fossil energy, 269
  coal (*see* Coal)
  natural gas, 10, 10*f*
    consumption, 303–304*t*
    demand for, 122–123, 123*t*, 123*f*
    disadvantaged geographical location, 124
    gas consumption for power generation, 126
    and gas-fired power generation, 56–57
    power generation technologies, 255
    prices of, 273, 274*t*
    production, 305–306*t*
    recovery rate of, 122–123
    resources and supply capability of, 125, 125*t*
    shale gas development, predicted output and
      importing of, 125, 126*t*
    supply of, 123, 123*t*
  oil, 118–122
Frequency modulation analysis model, 281–282,
    282*f*

G
Gas-fired power generation, 56–57
GESP-IV model. *See* Multiregional power source
    expansion and optimization model
    (GESP-IV)
Global population growth trend, 2, 2*f*
Gorges Hydropower Station, 85–86
Gross domestic product (GDP), 78–79, 128–129,
    219, 223, 270
Guandi hydropower station, 232, 233*f*, 234*t*
Guizui hydropower station, 234, 235*f*, 236*t*

H
Houziyan hydropower station, 232–233, 235*f*, 236*t*
Huangjinping hydropower station, 233–234, 235*f*,
    236*t*

Hydropower, 135–136
  bases, development of, 57
  development path and supporting conditions, 226*f*
    cascade hydropower stations, 230–235
    challenges, 227
    clean energy power generation, 227, 229
    hydropower projects, approval of, 235–237
    installed capacity, 227, 229–230, 234–235
    large-scale hydropower construction, capability of, 230
  development planning, 89–90
  development potential, 89
  ecological and environment impact assessment system, 244
  installed capacity of, 85–88, 87–88*f*
  international rivers, development planning of, 245
  project approval, annual target of, 245
  project evaluation, mechanism for, 245
  resettlement for, 244
  resources and distribution, 81–85
  self-use and transmission plan, 246
  slowdown scenario
    accelerated installed solar power capacity, 146
    accelerated installed wind power capacity, 145–146
    installed hydropower capacity, 145
  southwest hydropower, development and transmission of, 57–58
  technological innovation, 256

I

Import energy, 47–48
Input-output (IO) model, 66, 77–79, 215, 223
Integrated gasification combined cycle (IGCC) technology, 255
International Energy Agency (IEA), 40–41, 129–130

J

Jinchuan hydropower station, 232–233, 235*f*, 236*t*
Jinping stage-I hydropower station, 232, 233*f*, 234*t*
Jinping stage-II hydropower station, 232, 233*f*, 234*t*
Jinsha River, 230–231

K

Kala hydropower station, 232, 233*f*, 234*t*

L

Laoyingyan hydropower station, 233–234, 235*f*, 236*t*
Lead-acid cells, 228–229

Lenggu hydropower station, 232, 233*f*, 234*t*
Lianghekou hydropower station, 232, 233*f*, 234*t*
Lithium battery, 228–229
Liujiaxia Hydropower Station, 85–86
Longtoushi hydropower station, 233–234, 235*f*, 236*t*
Low power demand plan, 148
Luding hydropower station, 233–234, 235*f*, 236*t*

M

Mengdigou hydropower station, 232, 233*f*, 234*t*
Mercury emissions, 53
Micro-power grid technology, 259
Multiregional power source expansion and optimization model (GESP-IV), 275
  constraints, 277
  cost efficiency, 275
  functions, 275–276
  mixed integer programming, mathematical description of, 276–277
  output of, 277
  principle structure of, 278*f*
  target function, 276

N

National Bureau of Statistics of China
  energy indicators of, 269–271
  PECS data, 39, 39*t*
Natural gas, 10, 10*f*
  consumption, 126, 303–304*t*
  demand for, 122–123, 123*t*, 123*f*
  disadvantaged geographical location, 124
  and gas-fired power generation, 56–57
  power generation technologies, 255
  prices of, 273, 274*t*
  production, 305–306*t*
  recovery rate of, 122–123
  resources and supply capability of, 125, 125*t*
  shale gas development, predicted output and importing of, 125, 126*t*
  supply of, 123, 123*t*
New energy, 257, 271–272
Nitrogen oxide emissions, 52, 77
Non-fossil energy, 65–66, 269, 271
  benefits and costs of
    electric power supply, 190, 193–194*t*, 194*f*
    installed power capacity, 188
  clean energy power generation, 135
  development goals (*see* Development goals, non-fossil energy)
  electric power generation, 265
  end-user energy consumption structure, 42–43
  hydropower (*see* Hydropower)

large-scale development and utilization
of, 266
laws and regulations, 250–252
non-fossil energy mix, 188, 189–190*f*, 192*f*
nuclear power (*see* Nuclear power)
PECS
evolution of, 37–38, 38*f*
of major energy-consuming countries, 38–39,
39*t*
PEPS, 39
policy support system, 251, 268
power generation structure, 40–41
power grid development (*see* Power grids)
and power system, optimization and planning of
frequency regulation verification, analysis
process of, 71, 72*f*
maintenance scheduling and unit commitment,
improvements in, 69
multiregional power source optimization
planning process, 69, 70*f*
non-fossil energy sources, consumption and
proportion of, 72
production simulation and analysis process, 70
quota system, 69
research model, framework of, 67–69
proportion of, 185–188
renewable energy sources, 135
solar power (*see* Solar power)
sources, utilization of, 271
wind power (*see* Wind power)
Nonrenewable energy, 269
Nonwater renewable energy sources, 260–261, 267
North-South Power Transmission Project, 263–264
Nuclear energy resources
development planning, 96–97
development potential, 96
development situation
existing, under-construction and planned
nuclear stations, distribution of, 93, 95*f*
installed nuclear power capacity, changes of,
92, 93*t*
under-construction nuclear power projects, 93,
94*t*
uranium resources reserves and distribution,
90–92
Nuclear power
"actively develop nuclear power", principle of,
246
under construction, installed capacity of, 58
development path and supporting conditions,
226*f*
clean energy power generation, 227, 229
installed capacity, 227, 229

large-scale nuclear power development, 227,
237–239
nuclear power station construction, demands
for, 237
equipment design and manufacturing capability,
improvement of, 248
and gas power unit, techno-economic
parameters of, 272, 273*t*
investment and financial channels, expansion on,
247
laws and regulations, 246
nuclear fuel supply, capacity of, 248
power-generating fuel prices, 273, 274*t*
power plant site planning and protection, 247
strengthening top-level design, 247
talent team building, 248
technological innovation, 256
unified technical route, 247
uranium resources
availability of, 237
exploration technologies, development and
research of, 248
reserves and distribution, 90–92
supply, capacity of, 248

O
Offshore wind energy resources, 98–99
annual average wind power density,
distribution of, 99, 99*f*
potential development capacity, 97–99
water depth, 99
Oil resources
conventional oil resources, 119
import, 122
nonconventional oil resources, 120, 121*t*
recoverable reserves of, 118–122
Onshore wind energy resources, 97–99, 98*t*
Optimization
demand slowdown scenario
clean energy consumption and utilization,
181–182
coal power, 179
economical efficiency, 183–185
gas power stations, 180
overall system operation, 182
power flow, overall changes of, 181
pumped storage, 180
solar generation consumption and utilization,
183
wind power consumption and utilization, 183
expected development scenario
clean energy consumption and utilization,
154–156, 161–163

Optimization *(Continued)*
   coal power, 149–150
   cross-regional power flow, 153
   economical efficiency, 163–165
   gas power station, 151–152
   overall system operation, 156–157
   pumped storage, 150, 160–161
   solar power, operation and utilization of, 160
   wind power, operation and utilization of, 158–160
  hydropower slowdown scenario
   clean energy consumption and utilization, 168–169
   cross-regional power flow, 167–168
   economical efficiency, 171–173
   installed power capacity, 165–167
   overall system operation, 169
   solar power, operation and utilization of, 170–171
   wind power, operation and utilization of, 169–170
  nuclear power slowdown scenario
   clean energy consumption and utilization, 175
   economical efficiency, 176–178
   installed power capacity, 173–174
   overall system operation, 175
   power flow, overall changes of, 175
   solar power, operation and utilization of, 175–176
   wind power operation and utilization, 175
Overhaul schedule, 272

P

PECS. *See* Primary-energy consumption structure (PECS)
PEPS. *See* Primary energy production structure (PEPS)
Petroleum
  consumption, 299–300*t*
  production, 301–302*t*
Photovoltaic (PV) cell, 107
Photovoltaic (PV) power generation, 260
Physical energy moment, 285, 288*f*
Power-generating fuel prices, 273
Power generation structure, 40–41
Power grids, 228
  cross-regional UHV power transmissions, 197
  economic benefits
   effective installed capacity and power source investment, reduction of, 216
   electric power supply, reduction of, 216
   energy delivery, increase of, 216
   grid connection benefits, 216

   utilization efficiency, power grid equipment, 216
  energy-saving benefits
   coal consumption for power generation, reduction of, 218
   coal transportation and power transmission, 217
   electric power system, operation efficiency of, 218
   energy supply security, 217
   fossil energy consumption, reduction of, 217
   power grid transmission efficiency and lower line loss, 218
   terminal electric equipment, energy utilization efficiency of, 218
  energy structural adjustment and power grid functions, 201*f*
   electricity consumption, new energy development and services of, 199
   large-scale and long-distance energy and power transmission, 198
   power-receiving regions, energy sources in, 198
   unified power market, 200
  evaluation of, 214–215
  in future, 202–203
  hydropower base and power transmission, development of, 204
  integrated national energy, 198, 199*f*
  investments, 274
  large-scale development and new energy consumption
   distributed development and in situ utilization of new energy, 211
   "North China, East China, and Central China" UHV synchronous power grid, wind power, 207
   wind power, centralized development and long-distance transmission of, 206
  nationwide interconnection of, 50, 51*f*
  nuclear power, safe and economical operation of, 212–214
  power flow, 200–202
  receiving-end UHV synchronous power grids, 197
  resource and environmental benefits
   carbon dioxide emissions, reduction of, 219
   energy delivery and benefits of land use, 220
   environmental loss, reduction of, 219
   power industry PM 2.5 pollution, reduction of, 219
  safety benefits, 221–222
  social benefits, 223

technological innovation, 256
water curtailment reduction and increase
    hydropower utilization efficiency, 205
Power transmission, 60
    autonomous regions and municipalities,
        capability of, 274
    proportion of, 48–49
    social and economic benefits, 59–60
    trans-provincial and trans-regional power
        transmission capacity, 50
Power transmission and distribution (PTD), 18
Primary energy, 6–9, 7f
    consumption
        distribution of, 293–294t
        global total primary energy consumption
            growth trend, 3, 3f
        structure of, 291–292t
        total, 289–290t
    definition, 269
    nonrenewable energy, 269
    renewable energy, 269
    total energy production, 269–270
Primary-energy consumption structure (PECS)
    evolution of, 37–38, 38f
    of major energy-consuming countries, 38–39, 39t
    and PEPS, 39
Primary energy production structure (PEPS), 39
Pubugou hydropower station, 234, 235f, 236t
Pumped storage power stations, 259–260, 272–273,
        274t
    early-selected site region, total capacity of, 89,
        89t
    installed capacity of, 87–88, 88f
    techno-economic parameters of, 272–273, 274t

R
Random production simulation model
    analysis target, 279
    constraints, 279
    output results, 279
    principle structure of, 279, 279f
    structure and process of, 281f
    unit overhaul plan, structure and process of, 279,
        280f
Regional Investment Plans (RIP), 21
Renewable Energy Act 2009, 19

S
Sand oil resources, 120–121, 121t
Scenario analysis methods
    design
        conventional coal power, 140

cross-regional power grid expansion scale
    (power flow), 141
expected development scenario, 138–139,
    141–143
gas power station, 140–141
hydropower development slowdown scenario,
    138–139, 144–146
nuclear power development slowdown
    scenario, 138–139, 146–148
power demand slowdown scenario, 138–139,
    148
power grid zoning plan, 139
power source construction as input condition,
    139–140
pumped storage, 140
energy demands, predictions on
    comprehensive analysis, 131
    domestic organizations, 128–129
    international organizations, 129–130
    State Grid Energy Research Institute, 130–131
non-fossil energy development
    benefits and costs of, 188–190
    clean energy power generation, 135
    hydropower development, 135–136
    non-fossil energy mix, 188
    nuclear power, 135–136
    proportion of, 185–188
    renewable energy sources, 135
    solar power, 135, 137
    wind power, 135, 137
optimization
    demand slowdown scenario, 178–185
    expected development scenario, 148–165
    hydropower slowdown scenario, 165–173
    nuclear power slowdown scenario, 173–178
power demand
    economic situation, 132–133
    medium- and long-term, 131
    total electricity demand, 133–134, 134f
principles of
    cleanliness, 128
    economy, 127
    safety, 127
total supply of non-fossil energy, 138
Secondary energy, 269
"Second-generation" nuclear power technology,
    237–238
Self-balanced power development, 261
SGERI. See State Grid Energy Research Institute
    (SGERI)
Shale gas
    development, predicted output and importing of,
        125, 126t

Shale gas *(Continued)*
  exploitation, 12, 13*f*
  production, 13–14
Shale oil, 120–121, 121*t*
Shale oven technology, 121
Shaping hydropower station, 234, 235*f*, 236*t*
Shawan hydropower station, 234, 235*f*, 236*t*
Shenxigou hydropower station, 234, 235*f*, 236*t*
Shuangjiangkou hydropower station, 232–233,
    235*f*, 236*t*
Smart grid, 212, 213*f*, 252–253
Smart power grid technology, 259
Sodium-sulfur cells, 228–229
Solar energy resources
  annual solar radiation, 105–107, 106*t*
  development planning and policies, 108–109
  development potential, 107–108
  distribution of, 105, 105*f*
  installed solar power capacity, 107
  PV cell, 107
  solar thermal power generation technology, 107
Solar power, 135, 137, 259–260
  centralized and decentralized development, 59
  CSP development, 27–29, 28*f*
  development path and supporting conditions,
      226*f*
    grid connection and consumption, 239–240
    installed capacity of, 229
    photothermal power generation projects,
        demonstration of, 228
    PV power generation technology, 239
    rooftop grid connection systems, 228
  grid-connected operation of, 259–260
  technological innovation, 257
Solar thermal power generation technology, 107
State Grid Energy Research Institute (SGERI),
    73–75, 130–131
Sulfur dioxide emissions, 52, 77
Surplus electric power trade market, 18
Synchronous power grids, 203, 203*f*

T

Techno-economic parameters, for power source
      optimization
  coal power unit, 272, 273*t*
  nuclear and gas power unit, 272, 273*t*
  power-generating fuel prices, 273
  pumped storage power stations, 272–273, 274*t*
  unit reserve and overhaul schedule, 272
"Third-generation" nuclear power technology,
    237–238
"Three Gorges" project, 244
Tongjiezi hydropower station, 234, 235*f*, 236*t*

Tongzilin hydropower station, 232, 233*f*, 234*t*
Total energy consumption, 267–268, 270
  control measures, 241–243
  CTEC, 54–55
  energy loss, 270
  final energy consumption, 270
  loss from energy processing and conversion, 270
Total energy production, 269–270

U

Ultrahigh voltage (UHV), 197–200, 214, 219–221,
    228, 261–262
Uranium mines, 90–91, 92*t*
Uranium resources
  availability of, 237
  exploration technologies, development and
      research of, 248
  reserves and distribution, 90–92
  supply, capacity of, 248
US Energy Independence Strategy, 10
  basic conditions, 11–12, 11*t*
  clean energy generation, 14
  global energy balance by regions, 13
  quantity and price controllability, 13
  shale gas exploitation, 12, 13*f*
  shale gas production, 13–14

V

Virtual power plant, 18

W

West-East Power Transmission project, 85–86,
    263–264
Wind energy resources
  development planning and policies, 103–104
  distribution of, 97, 98*f*
  exploration and evaluation, 97
  installed grid-connected wind power capacity
      and generation, 100–101, 101*f*, 101*t*
  key regions, evaluation in, 99–100, 100*t*
  observation, professional network for, 97
  offshore wind energy resources, 97–99, 99*f*
  onshore wind energy resources, 97–99, 98*t*
  wind power curtailment, 102, 102*t*
  wind power development potential, 102–103
Wind power, 135, 137, 259–260
  average annual growth rate of, 49
  centralized and decentralized development,
      58–59
  centralized and distributed development,
      equal emphasis on, 249
  development path and supporting conditions,
      226*f*

grid connection and consumption, 239–240
  installed capacity of, 229, 239
  technical feasibility, problems of, 227–228
  technological level and reliability of, 239
distributed mode, 27, 28f
grid-connected operation of, 259–260
innovations in, 268
installed capacity of, 49–50
offshore wind power in Europe, 29, 29t
peak shaving, compensation mechanism for, 249
power transmission corridors, construction of,
  249
technological innovation, 257
World energy consumption and production
  coal, 9, 9f, 295–298t
  natural gas, 10, 10f, 303–306t
  oil, 9–10, 9f, 299–302t
  primary energy, 6–9, 7f, 289–292t
  world EUECS, 6, 7t
World energy development
  development experience
    fully leverage flexible power regulation
      capacity, 21–22
    large-scale integration and consumption,
      through interconnected power grid, 22–23
    orderly guidance leading, clean energy, 19–20,
      20t
    power generation and transmission, 20–21, 21f
  development trend
    electric power, 32–36, 33f, 35f
    energy storage technology, 29–31, 30–31f
    European-American power grid integration
      (see European-American power grid
      integration)
    wind power and solar power, 27–29, 28f, 29t
  EU Energy Strategy, 14
    2030 Climate and Energy Policy Framework,
      14–16

Energy 2020, 14–15
  energy development goal, 14–15, 15t
  Energy Roadmap 2050, 14, 16
features, 31–32, 32t
Japan's energy strategy
  Energy Reform Strategy, 16–18
  postearthquake denuclearization plan and
    energy strategy adjustment, 16–17
low-carbon green development, 1
population, economy, environment, and energy
  consumption, 2–5, 2–4f
renewable energy development goals, 18, 19t
transformational stages, 1
US Energy Independence Strategy, 10
  basic conditions, 11–12, 11t
  clean energy generation, 14
  global energy balance by regions, 13
  quantity and price controllability, 13
  shale gas exploitation, 12, 13f
  shale gas production, 13–14
World primary energy consumption
  by regions, 6–9, 7f
  structure, 5–6, 6f
Wudongde hydropower station, 231, 231f, 232t

X

Xiangjiaba hydropower station, 231, 231f, 232t
Xi Luodu hydropower station, 231, 231f, 232t

Y

Yagen hydropower station, 232, 233f, 234t
Yalong River Basin, 232
Yangfanggou hydropower station, 232, 233f, 234t
Yingliangbao hydropower station, 233–234, 235f,
  236t

Z

Zhentouba hydropower station, 234, 235f, 236t

Printed and bound by CPI Group (UK) Ltd, Croydon, CR0 4YY

08/05/2025

01864802-0001